Engineers and applied geophysicists routinely encounter interpolation and estimation problems when analyzing data from field observations. *Introduction to Geostatistics* presents practical techniques for the estimation of spatial functions from sparse data. The author's unique approach is a synthesis of classic and geostatistical methods, with a focus on the most practical linear minimum-variance estimation methods, and includes suggestions on how to test and extend the applicability of such methods.

The author includes many useful methods often not covered in other geostatistics books, such as estimating variogram parameters, evaluating the need for a variable mean, parameter estimation and model testing in complex cases (e.g., anisotropy, variable mean, and multiple variables), and using information from deterministic mathematical models.

Well illustrated with exercises and worked examples taken from hydrogeology, *Introduction to Geostatistics* assumes no background in statistics and is suitable for graduate-level courses in earth sciences, hydrology, and environmental engineering and also for self-study.

INTRODUCTION TO GEOSTATISTICS

STANFORD–CAMBRIDGE PROGRAM

The Stanford–Cambridge Program is an innovative publishing venture resulting from the collaboration between Cambridge University Press and Stanford University and its Press.

The Program provides a new international imprint for the teaching and communication of pure and applied sciences. Drawing on Stanford's eminent faculty and associated institutions, books within the Program reflect the high quality of teaching and research at Stanford University.

The Program includes textbooks at undergraduate and graduate level, and research monographs, across a broad range of the sciences.

Cambridge University Press publishes and distributes books in the Stanford–Cambridge Program throughout the world.

INTRODUCTION TO GEOSTATISTICS: Applications to Hydrogeology

P. K. KITANIDIS

Stanford University

PUBLISHED BY THE PRESS SYNDICATE OF THE UNIVERSITY OF CAMBRIDGE
The Pitt Building, Trumpington Street, Cambridge, United Kingdom

CAMBRIDGE UNIVERSITY PRESS
The Edinburgh Building, Cambridge CB2 2RU, UK http: //www.cup.cam.ac.uk
40 West 20th Street, New York, NY 10011-4211, USA http: //www.cup.org
10 Stamford Road, Oakleigh, Melbourne 3166, Australia

First published 1997
Reprinted 1999

A catalogue record for this book is available from the British Library

Library of Congress Cataloguing-in-Publication data

Kitanidis, P. K. (Peter K.)
Introduction to geostatistics : applications to hydrogeology /
P. K. Kitanidis
p. cm.
Includes bibliographical references
1. Hydrogeology – Statistical methods. I. Title
GB1001.72.S7K57 1997
551.49'072–dc20 96-28608
CIP

ISBN 0 521 58312 8 hardback
ISBN 0 521 58747 6 paperback

Transferred to digital printing 2003

I dedicate the book to the fond memory of my Parents.

Contents

List of tables

List of figures

Preface

This book grew out of class notes for a course that served two purposes:

1. To familiarize graduate students in hydrogeology and environmental engineering with some practical methods for solving interpolation and related estimation problems. The course emphasized geostatistical methods.
2. To illustrate how one may use data to develop empirical models, that is, to provide an introduction to applied statistical modeling.

Engineers and applied geophysicists routinely encounter estimation problems: From data, they must surmise the values of unknown variables. A case in point is drawing the map of the free surface (elevation of water table) in an aquifer from observations of the free-surface elevation at a few wells in addition to other information. A practitioner faces such estimation problems much more often than some other problems that are covered in the academic curriculum. And yet most practitioners have received no formal instruction on estimation and, in many cases, are unable to judge the applicability of a method or to interpret the results produced by software they use. Their efforts to develop a grasp of the subject are frustrated by unfamiliar jargon. Indeed, the type of training one receives in mathematics and physics does not help one to develop the skills needed for using data to build empirical models. I believe that it is absurd to expect one to "pick up" estimation methods without some systematic training and a fair degree of effort.

After teaching the course for about ten years, I realized that there might be room for a textbook that presumes no background in statistics and that uses common sense to motivate and justify methods used for estimation of spatial functions. This book resulted from this realization. As it propounds methods and tools that the practitioner is likely to use, the book discusses in plain terms the reasoning involved in building up empirical models and fitting parameters. Jargon and mathematical abstraction have been avoided as much as possible. Nevertheless, the student is expected to have had a calculus-based course in probability theory and to have at least a rudimentary knowledge of linear algebra.

The book could have been much shorter if a more abstract approach had been followed. However, to write a single equation that describes ten different applications does not mean that one has understood at once all these applications! To proceed from the general to the specific is mathematically elegant but more appropriate for advanced texts, because it requires some degree of familiarity with the methods. For an introductory textbook, particularly on a subject so foreign to the intended audience, my experience has taught me that the only approach that works is to proceed from the simple and specific to the more complex and general. The same concepts are discussed several times, every time digging a bit deeper into their meaning.

Because statistical methods rely to a great extent on logical arguments it is particularly important to study the book from the beginning. Although this book may appear to be full of equations, it is not mathematically difficult provided again that one starts from the beginning and becomes familiar with the notation. The book is intended for a one-semester course for graduate-level engineers and geophysicists and also can be used for self-study. The material is limited to linear estimation methods: That is, we presume that the only statistics available are mean values and covariances. I cannot overemphasize the point that the book was never meant to be a comprehensive review of available methods or an assessment of the state of the art.

Every effort has been made to catch mistakes and typographical errors, but for those that are found after the publication of the book, a list of errata will be maintained at

http://www-ce.stanford.edu/cive/faculty/Kitanidis.html

I thank all my co-workers and students at Stanford and elsewhere who with their comments have assisted me so much in writing this book.

Peter K. Kitanidis
Palo Alto, California
September 1996

1
Introduction

1.1 Introduction

It is difficult and expensive to collect the field observations that an environmental engineer or hydrogeologist needs to answer an engineering or scientific question. Therefore, one must make the best use of available data to estimate the needed parameters. For example, a large number of measurements are collected in the characterization of a hazardous-waste site: water-surface level in wells, transmissivity and storativity (from well tests), conductivity from permeameter tests or borehole flowmeters, chemical concentrations measured from water and soil samples, soil gas surveys, and others. However, because most subsurface environments are complex, even a plethora of data is not sufficient to resolve with accuracy the distribution of the properties that govern the rates of flow, the rates of transport and transformation of chemicals, or the distribution of concentrations in the water and the soil. The professionals who analyze the data must fill in the gaps using their understanding of the geologic environment and of the flow, transport, or fate mechanisms that govern the distribution of chemicals.

However, process understanding is itself incomplete and cannot produce a unique or precise answer. Statistical estimation methods complement process understanding and can bring one closer to an answer that is useful in making rational decisions. Their main contribution is that they suggest how to weigh the data to compute best estimates and error bounds on these estimates. Statistics has been aptly described as a guide to the unknown; it is an approach for utilizing observations to make inferences about an unmeasured quantity. Rather than the application of cookbook procedures, *statistics is a rational methodology* to solve practical problems. The purpose of this book is to provide some insights into this methodology while describing tools useful in solving estimation problems encountered in practice. Two examples of such problems are: *point estimation* and *averaging*.

In point estimation one uses measurements of a variable at certain points to estimate the value of the same variable at another point. For example, consider measurements of concentration from the chemical analysis of soil samples from borings. The question is how to estimate the concentration at the many other locations where soil samples are unavailable. Another example is drawing lines of constant transmissivity (in other words, contour lines of transmissivity) from the results of pumping tests at a number of nonuniformly spaced wells. Drawing a contour map is equivalent to interpolating the values of the transmissivity on a fine mesh. Examples can be found in references [54, 55, 105, 106, 119, 116, 141, and 15].

In averaging one uses point estimates of concentration to determine the average concentration over a volume of soil; this estimate is needed for evaluation of the total mass of contaminants. Another example, drawn from surface hydrology, is the estimation of mean areal precipitation over a watershed from measurements of rainfall at a number of rain gages.

Due to complexity in the spatial variability of the variables involved, one cannot obtain exact or error-free estimates of the unknowns. Statistical methods weigh the evidence to compute *best estimates* as well as *error bars* that describe the potential magnitude of the estimation error. Error bars, or information about how much faith to put in the estimates, are essential in making engineering decisions. Statistical methods are applied with increased frequency to evaluate compliance with regulatory requirements because the best one can do is to provide a reasonable degree of assurance that certain criteria have been met. Also, using statistics one can anticipate the impact of additional measurements on error reduction before the measurements are taken. Thus, statistics is useful in deciding whether the present data base is adequate for detecting all important sources of contamination and, if not, where to collect the additional measurements so that the objectives of *monitoring* (such as demonstrating regulatory compliance) are met in the most cost-effective way.

Once one masters the application of the statistical methodology to relatively simple problems, such as those above, one can tackle more complicated problems such as estimating one variable from measurements of another. It is often convenient to use a variable that can be easily observed or computed to estimate another variable that is difficult or expensive to measure. For examples, (*a*) land topography may be used to estimate the phreatic surface elevation of a surficial aquifer; (*b*) overburden and aquifer thickness may correlate and can be used to estimate the transmissivity of a confined permeable layer; and (*c*) hydraulic head measurements may provide information about the transmissivity and vice versa.

Such problems can be rather challenging, particularly if one integrates data analysis with mathematical models that simulate geologic, flow, or transport processes. This book deals with relatively simple applications, but the same general methodology applies to complicated problems as well. In fact, the power of the methods to be described becomes most useful when utilizing measurements of different types, combining these with deterministic flow and transport models, and incorporating geological information to achieve the best characterization possible. However, one is advised not to attempt to solve complex problems before developing a sound understanding of statistical techniques, which is obtained only through practice starting with simpler problems. Statistical methods are sometimes misapplied because professionals who use them have received no training and apply them without awareness of the implicit assumptions or a firm grasp of the meaning of the results. (It has been said that they are often used the way a drunk uses a lamppost: for support rather than illumination.) Blindly following methods one does not understand provides countless opportunities for misapplication.

1.2 A simple example

This first example will be used to provide context and motivate the application of some of the techniques that will follow. (To understand all the details, some readers will find it useful to go through the review of basic probability theory presented in Appendix A.)

Well tests were conducted at eight wells screened in a confined aquifer providing values of the transmissivity. (The location of the wells on plan view is shown as o in Figure 1.1. The values are given in Table 1.1.) The question is:

> Given the information currently available, if a well were drilled at another location (indicated by an × in Figure 1.1) and a similar pumping test were conducted, what value of the transmissivity would be observed?

Table 1.1. *Transmissivity data for example in this section*

T (m²/day)	x (km)	y (km)
2.9	0.876	0.138
2.5	0.188	0.214
4.7	2.716	2.119
4.2	2.717	2.685
4.2	3.739	0.031
2.1	1.534	1.534
2.4	2.078	0.267
5.8	3.324	1.670

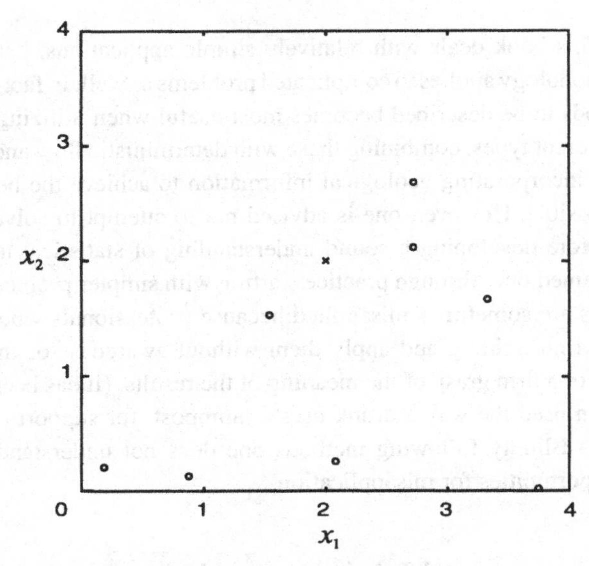

Figure 1.1 Location of transmissivity measurements (○) and unknown (×).

Assume also that the driller's logs indicate that all wells were drilled in the same formation and geologic environment. There are good reasons to believe that the formation is a confined aquifer bounded by nearly impermeable layers above and below. Beyond that, however, and despite considerable geological information available at the site, the variability among transmissivity measurements cannot be explained in terms of other measurable quantities in a manner useful in extrapolation to unmeasured locations. If we actually admit that we cannot explain this variability, how can we extrapolate from the sample of the eight observations to the unknown value? The point is that, because we cannot come up with a deterministic mechanism that explains variability, we postulate a *probabilistic model*, i.e., a set of mathematical equations that summarize what we know and are useful in making predictions.

The simplest approach is to compute the frequency distribution of the data and then to use it to describe the odds that the transmissivity at the location of interest will have a certain value. The premise is that "each transmissivity observation is randomly and independently sampled from the same probability distribution." It is like saying that every time we measure the transmissivity we perform an experiment whose outcome is a random variable with given probability distribution. Of course, this experiment is only a convenient concept; this simple model is not meant to represent the physical reality of what transmissivity is or how it is measured, but rather, it constitutes a practical and reasonable way to use what we know in order to make predictions.

We are still faced with the problem of estimating the probability distribution. We may approximate it with the experimental probability distribution (i.e., the distribution of the data); that is, we assume that the transmissivity takes any of the eight observed values with equal probability 1/8. Now we are able to make predictions. According to the model that we constructed, we predict that the value of transmissivity at point x is $T_1, T_2, \ldots,$ or T_8 with equal probability 1/8.

Such a model may appear crude, but it is a rational way to use experience as a guide to the unknown. In fact, this simple model is adequate for some practical applications. Also, the reader should not be left with the impression that the approach boils down to subjective judgement. The questions of the validity or suitability of a model and of the sensitivity of the prediction to modeling assumptions can be addressed, but this is left for later in this book.

In many applications, what is needed is to determine a good estimate of the unknown, $\hat{T}_0$, and a measure of the error. An *estimator* is a procedure to compute $\hat{T}_0$ from the data. Even though we cannot foretell the actual error in the estimate, we can describe its probability distribution. It is common to measure the anticipated error by the quantity known as the *mean square error*, i.e., the expected value (average weighted by the probability) of the square difference of the estimate from the true value:

$$\sigma_0^2 = \frac{1}{8}(\hat{T}_0 - T_1)^2 + \frac{1}{8}(\hat{T}_0 - T_2)^2 + \cdots + \frac{1}{8}(\hat{T}_0 - T_8)^2. \tag{1.1}$$

After some algebraic manipulation to rearrange terms, the expression for the mean square error becomes

$$\sigma_0^2 = \left(\hat{T}_0 - \frac{T_1 + T_2 + \cdots + T_8}{8}\right)^2 + \frac{1}{8}(T_1^2 + T_2^2 + \cdots + T_8^2) - \left(\frac{T_1 + T_2 + \cdots + T_8}{8}\right)^2. \tag{1.2}$$

Equation (1.2) demonstrates that the value of $\hat{T}_0$ that makes the mean square error as small as possible is the arithmetic mean of the observations,

$$\hat{T}_0 = \frac{T_1 + T_2 + \cdots + T_8}{8}, \tag{1.3}$$

which is the estimate with mean square error

$$\sigma_0^2 = \frac{1}{8}(T_1^2 + T_2^2 + \cdots + T_8^2) - \left(\frac{T_1 + T_2 + \cdots + T_8}{8}\right)^2 \tag{1.4}$$

An estimator with minimum mean square error will be referred to as a *best* or *minimum-variance* estimator.

Using Equation (1.3), the expected value of the estimation error is

$$\frac{1}{8}\left(\frac{T_1 + T_2 + \cdots + T_8}{8} - T_1\right) + \frac{1}{8}\left(\frac{T_1 + T_2 + \cdots + T_8}{8} - T_2\right)$$
$$+ \cdots + \frac{1}{8}\left(\frac{T_1 + T_2 + \cdots + T_8}{8} - T_8\right) = 0. \tag{1.5}$$

When the expected value of the estimation error is zero, the estimator is called *unbiased*.

Notice that the estimate was obtained from a formula that looks like

$$\hat{T}_0 = \lambda_1 T_1 + \lambda_2 T_2 + \cdots + \lambda_n T_n, \tag{1.6}$$

which, using the summation sign notation, can be written as

$$\hat{T}_0 = \sum_{i=1}^{n} \lambda_i T_i, \tag{1.7}$$

where n is the number of measurements and $\lambda_1, \lambda_2, \ldots, \lambda_n$ are coefficients or weights. In this example, the weight λ_i can be conveniently interpreted as the probability that the unknown equals the value of the i-th measurement, assuming that the only possible values are the observed ones. This expression, Equation (1.7), is known as a *linear estimator*.

In practice, the most useful class of estimators comprises best (minimum-variance) linear unbiased estimators (affectionately known as *BLUEs*), which is the subject of this book.

As already mentioned, for any estimator, the error is a random variable, i.e., it is described through a probability distribution. In the example, the error is $(\hat{T}_0 - T_i)$, for i from 1 to 8, with equal probability 1/8. If these errors are symmetrically distributed about a central value, as are those of Figure 1.2, and

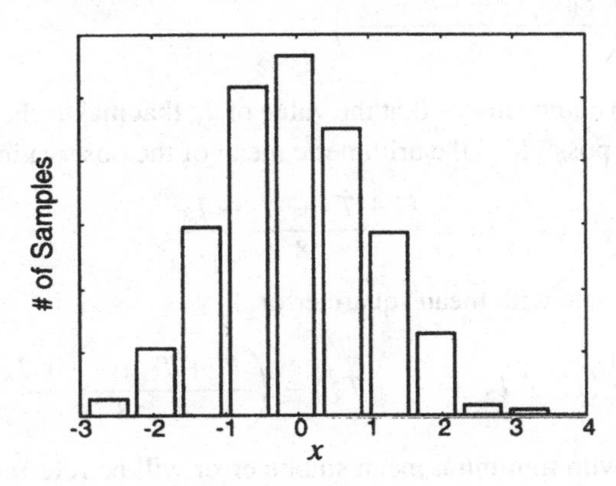

Figure 1.2 Distribution of nearly normal data.

follow a bell-shaped distribution that resembles a *normal* (*Gaussian*) distribution, then the mean of the distribution is the most representative value and the mean square value is the definite measure of the spread. However, in the case of transmissivities or hydraulic conductivities and concentrations, the histogram of these errors usually indicates that the distribution is not symmetric. In this case, there is no unequivocal representative value. In addition to the mean, there are other possibilities, such as the value that minimizes the mean absolute error or the median (the value exceeded by half the values). For this reason, the minimum-variance estimators are most suitable when we have reason to believe that the frequency distribution of the estimation errors may resemble the normal distribution. In the case of transmissivities or concentrations, it is common to use linear estimators with the logarithm of transmissivity, $Y = \ln(T)$, instead of with T. That is, instead of analyzing T we analyze Y. Examples will be seen in other chapters.

The weights in the linear estimator used in this example are equal; that is, the probability that the unknown is equal to a measured value is presumed to be the same no matter how far or in what direction the unknown is located from the location of the observation. Also, the locations of the other measurements have no effect in the selection of the weight. In many situations, however, the transmissivity varies gradually in space in such a way that it is more likely that the unknown will resemble an observation near than an observation far from its location. Therefore, the weights should be nonuniform (larger for nearby observations). This book describes methods that analyze data for clues on how to compute these weights in a way that reflects the spatial variability of the quantity of interest as well as the location of measurements and the unknown.

1.3 Statistics

First, let us clear up a common misunderstanding. The word statistics (plural) means averages of numerical data, such as the batting average of a player or the median of a batch of hydraulic conductivity measurements. However, data can be misleading, when improperly analyzed and presented. The word statistics (singular) refers to a *methodology for the organization, analysis, and presentation of data*. In particular,

> *statistical modeling* is an approach for fitting mathematical equations to data in order to predict the values of unknown quantities from measurements.

Hydrogeologists and environmental and petroleum engineers, like scientists and engineers everywhere, use such methods on an almost daily basis so that some knowledge of statistics is essential today. Those who casually dismiss statistical methods are the ones most likely to misuse them or to be misled by them.

Basically, we are concerned with estimation problems in which the value of an unknown needs to be inferred from a set of data. It is convenient to subdivide the methodology into the following steps:

- Postulate a model that describes the variability of the data and can be used to extrapolate from the data to the unknown.
- If the model involves undetermined parameters, these parameters are estimated or *fitted*.
- The model is put to the test or *validated*.
- If the model is deemed acceptable, it is used to predict the unknown.

The postulated model is probabilistic. Parameter fitting, model validation, and prediction involve computations of probability distributions or moments (such as the mean, the variance, etc.). These models must be reasonably simple or else the computations may be just too complicated for the approach to be of practical use. The computations in the methods presented in this book are reasonably simple and involve only mean values, variances, and correlation coefficients. However, there are even more important reasons for selecting simple probabilistic models, as will be discussed later.

Conceptually, the part that novices in statistics have the most trouble understanding is the selection of the empirical model, i.e., the model that is introduced to fit the data. So let us say a few things on this subject. How do we know that we have the right model? The truth is that one cannot (and may not even need to) prove that the postulated model is the right one, no matter how many the observations. There is nothing that anyone can do about this basic fact, which is not a limitation of statistics but to various degrees affects all sciences that rely on empirical data. In the example of Section 1.2, we cannot prove the assumption that the data were somehow generated randomly from the same distribution and even more we cannot prove that the unknown was generated from the same distribution. However,

1. unless there is evidence to the contrary, it is not an unreasonable assumption, and
2. one can check whether the data discredit the assumption.

It is best to approach empirical models from a utilitarian perspective and see them as a practical means to:

1. summarize past experience and
2. find patterns that may help us to extrapolate.

Table 1.2. *Porosity versus location (depth)*

η	0.39	0.41	0.39	0.37	0.36	0.29	0.38	0.34	0.31	0.28	0.32	0.30
x(ft)	−24	−26	−28	−30	−32	−34	−36	−38	−40	−42	−44	−46

The model of Section 1.2 is another way of saying that frequency analysis is a practical way to describe the variability in the data and to make use of past observations in predicting future ones. It is a reasonable approach, which should lead to rational decisions.

A model should be judged on the basis of information that is available at the time when the model is constructed. Thus, a model that looks right with 10 measurements may be rejected in favor of another model when 100 measurements have been collected. It will be seen that, in all cases, the simplest empirical model consistent with the data is likely to be best for estimation purposes (a principle known as *Occam's razor*). Furthermore, it will be seen that one of the most important practical contributions of statistical estimation methods is to highlight the fine distinction between fitting (a model to the data) and obtaining a model that we may trust to some degree for making predictions.

Exercise 1.1 *What are the basic characteristics of an estimation problem? Describe an estimation problem with which you are familiar.*

Exercise 1.2 *Describe a common-sense approach to utilize observations (known facts) to make extrapolations or predictions (about unknown facts). Describe two examples, one from your everyday life and experience and one from scientific research (e.g., put yourselves in Newton's shoes and try to imagine how he came up with the basic law of motion $F = ma$). Outline the steps you follow in a systematic way. (You may find it useful to review what is known as the scientific method and discuss its generality and relevance to everyday life.)*

Exercise 1.3 *Consider observations of porosity in a borehole (first column is measured porosity, second column is depth) as shown in Table 1.2. Find the best estimate and standard error*[1] *at locations $x = -37$ ft using the simple model of Section 1.2. What is the significance of the standard error? Discuss the pros and cons of this simple model and whether it seems that this model is a reasonable description for this data set.*

[1] The standard error of estimation is the square root of the mean square error of estimation.

1.4 Geostatistics

In applied statistical modeling (including regression and time-series) least squares or linear estimation is the most widely used approach. Matheron [94 and 95] and his co-workers advanced an adaptation of such methods that is well suited to the solution of estimation problems involving quantities that vary in space. Examples of such quantities are conductivity, hydraulic head, and solute concentration. This approach is known as the *theory of regionalized variables* or simply *geostatistics*. Popularized in mining engineering in the 1970s, it is now used in all fields of earth science and engineering, particularly in the hydrologic and environmental fields. This book is an introductory text to geostatistical linear estimation methods.

The geostatistical school has made important contributions to the linear estimation of spatial variables, including the popularizing of the variogram and the generalized covariance function. Geostatistics is well accepted among practitioners because it is a down-to-earth approach to solving problems encountered in practice using statistical concepts that were previously considered recondite. The approach is described in books such as references [24, 36, 37, 70, 73, 121, and 76] with applications mainly in mining engineering, petroleum engineering, and geology. Articles on geostatistics in hydrology and hydrogeology include [7 and 102] and chapters can be found in [13 and 41]. A book on spatial statistics is [30]. Software can be found in references [50, 145, and 43] and trends in research can be discerned in reference [44].

The approach presented in this book departs from that of the books cited earlier (which, for the sake of convenience will be called "mining geostatistics") in consequential ways. For the readers who are already familiar with mining geostatistics, here is a list of the most important differences:

1. The estimation of the variogram in mining geostatistics revolves around the experimental variogram; sometimes, the variogram is selected solely on the basis that it fits the experimental variogram. This approach is simple to apply but unsatisfactory in most other aspects. In contrast, in the approach followed in this book, the variogram is selected so that it fits the data, i.e., the approach relies more on the minimization of a criterion of agreement between the data and the predictions than on the experimental variogram.
2. Unlike mining geostatistics, which again relies on the experimental variogram to select the geostatistical model, the approach preferred in this work is to apply an iterative three-step approach involving: 1. exploratory analysis that suggests a model; 2. parameter estimation; and 3. model validation, which may show the way to a better model. Model validation is implemented differently and has a much more important role than in mining geostatistics.

3. Ordinary kriging, which describes spatial variability only through a variogram and is the most popular method in mining geostatistics, can lead to large mean square errors of estimation. In many environmental applications, one may be able to develop better predictive models by judiciously describing some of the "more structured" or "large-scale" variability through drift functions. The error bars can be further reduced by making use of additional information, such as from the modeling of the processes. This additional information can be introduced in a number of ways, some of which will be seen in this book.

1.5 Key points of Chapter 1

This book is a primer of geostatistical estimation methods with applications in contaminant hydrogeology. Statistics is a methodology for utilizing data and other information to make inferences about unmeasured quantities. Statistical methods complement deterministic process understanding to provide estimates and error bars that are useful in making engineering decisions. The methods in this book are an adaptation and extension of linear geostatistics.

2
Exploratory data analysis

The analysis of data typically starts by plotting the data and calculating statistics that describe important characteristics of the sample. We perform such an exploratory analysis to:

1. familiarize ourselves with the data and
2. detect patterns of regularity.

Graphical methods are useful to portray the distribution of the observations and their spatial structure. Many graphical methods are available and even more can be found and tailored to a specific application. The modest objective of this chapter is to review common tools of frequency analysis as well as the experimental variogram. Exploratory analysis is really a precursor to statistical analysis.

2.1 Exploratory analysis scope

Before computers, hydrogeologists used to spend hours transcribing and plotting their data. Although time consuming, labor intensive, and subject to human errors, one cannot deny that this process enhanced familiarity with data to the extent that the analyst could often discern patterns or spot "peculiar" measurements. This intimacy with one's data might appear lost now, a casualty of the electronic transcription of data and the extensive use of statistical computer packages that perform the computations.

However, data analysis and interpretation cannot be completely automated, particularly when making crucial modeling choices. The analyst must use judgment and make decisions that require familiarity with the data, the site, and the questions that need to be answered. It takes effort to become familiar with data sets that are often voluminous and describe complex sites or processes. Instead of striving for blind automation, one should take advantage of available computers and computer graphics to organize and display data in ways unimaginable using manual methods (for review of basic ideas see, for example, [20]).

Measurements may vary over a wide range. In most cases it is impossible, and often useless, for any person to remember every single measurement individually. One may start by summarizing in a convenient way the behavior of measurements that act similarly and by pointing out the measurements that behave differently from the bulk of the data. What is the best way to organize and display the data? What are the measures that summarize the behavior of a bunch of data? And could it be that a certain data transformation can simplify the task of summarizing the average behavior of a data batch? These are some of the issues to be discussed. But first, here are three basic principles:

- It does little good to just look at tabulated data. However, the human eye can recognize patterns from graphical displays of the data.
- It is important to look at the data in many ways and to keep in mind that some techniques implicitly favor a certain model. During exploratory data analysis one should make as few assumptions as possible.
- Conclusions made during exploratory analysis are tentative. A model cannot be accepted on the basis of exploratory analysis only but should be corroborated or tested.

To illustrate the objectives and usage of exploratory data analysis, consider the following data sets:

1. Measurements of transmissivity of a thin sand-and-gravel aquifer at 8 locations (see Table 1.1 and Figure 1.1).
2. Measurements of potentiometric head at 29 locations in a regional confined sandstone aquifer (see Table 2.1 and Figure 2.1).

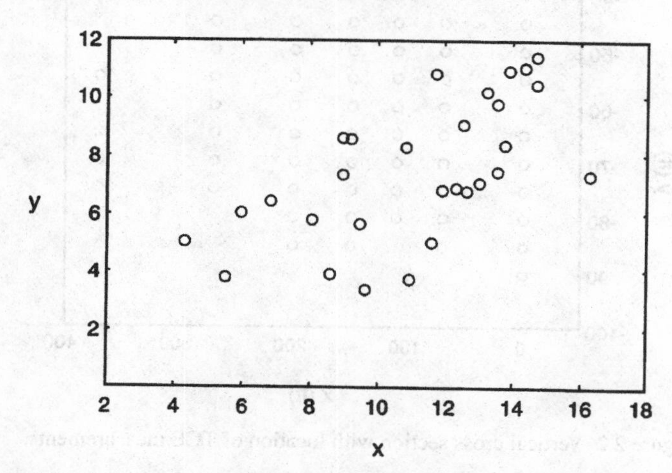

Figure 2.1 Plan view of aquifer showing location of head data.

3. Measurements at 56 locations of concentration of trichloroethylene (TCE) in groundwater on a transect in a fine-sand surficial aquifer (see Table 2.2 and Figure 2.2).

Table 2.1. *Head observations in a regional confined aquifer*

Head (ft)	x	y	Head (ft)	x	y
1061	6.86	6.41	662	13.23	10.18
1194	4.32	5.02	685	13.56	9.74
1117	5.98	6.01	1023	8.06	5.76
880	11.61	4.99	998	10.95	3.72
1202	5.52	3.79	584	14.71	11.41
757	10.87	8.27	611	16.27	7.27
1038	8.61	3.92	847	12.33	6.87
817	12.64	6.77	745	13.01	7.05
630	14.70	10.43	725	13.56	7.42
617	13.91	10.91	688	13.76	8.35
986	9.47	5.62	676	12.54	9.04
625	14.36	11.03	768	8.97	8.6
840	8.99	7.31	782	9.22	8.55
847	11.93	6.78	1022	9.64	3.38
645	11.75	10.8			

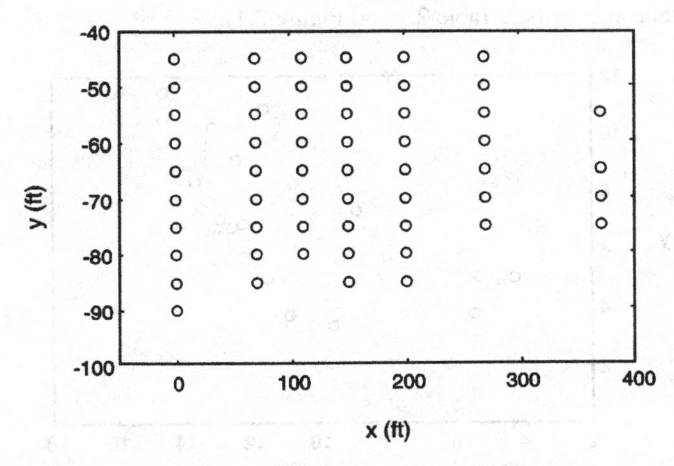

Figure 2.2 Vertical cross section with location of TCE measurements.

Table 2.2. *TCE concentrations in groundwater in a vertical cross section*

No	TCE (ppb)	x (ft)	y (ft)	No	TCE (ppb)	x (ft)	y (ft)
1	1.00e+01	0	−45	38	1.16e+03	200	−50
2	1.00e+01	0	−50	39	4.27e+03	200	−55
3	6.40e+01	0	−55	40	9.87e+03	200	−60
4	5.42e+02	0	−60	41	7.32e+03	200	−65
5	3.46e+02	0	−65	42	4.03e+03	200	−70
6	1.91e+02	0	−70	43	2.73e+02	200	−75
7	8.50e+01	0	−75	44	1.90e+02	200	−80
8	3.00e+01	0	−80	45	5.22e+02	200	−85
9	7.40e+00	0	−85	46	6.70e+01	270	−45
10	1.46e+01	0	−90	47	5.50e+01	270	−50
11	1.20e+01	70	−45	48	1.36e+02	270	−55
12	2.30e+01	70	−50	49	7.08e+02	270	−60
13	1.87e+02	70	−55	50	8.82e+02	270	−65
14	1.64e+02	70	−60	51	9.39e+02	270	−70
15	7.01e+02	70	−65	52	3.08e+02	270	−75
16	2.13e+04	70	−70	53	1.40e+01	370	−55
17	1.86e+04	70	−75	54	7.00e+00	370	−65
18	6.22e+02	70	−80	55	2.50e+00	370	−70
19	1.39e+03	70	−85	56	2.50e+00	370	−75
20	4.00e+01	110	−45				
21	4.20e+01	110	−50				
22	6.55e+02	110	−55				
23	2.16e+04	110	−60				
24	6.77e+04	110	−65				
25	3.89e+04	110	−70				
26	5.84e+02	110	−75				
27	2.54e+03	110	−80				
28	6.00e+00	150	−45				
29	6.63e+02	150	−50				
30	3.75e+03	150	−55				
31	8.76e+03	150	−60				
32	1.40e+04	150	−65				
33	1.61e+04	150	−70				
34	1.28e+04	150	−75				
35	7.63e+02	150	−80				
36	2.89e+02	150	−85				
37	1.14e+01	200	−45				

2.2 Experimental distribution

We want to describe the frequency distribution of a set of n measurements (such as those on Tables 1.1, 2.1, and 2.2) without regard to their location. We call this distribution "experimental" or "empirical" because it depends only on the data.

We can describe the distribution of a set of data through the histogram, the ogive, and the box plot. We will also review numbers that represent important characteristics of a data set, such as central value, spread, and degree of asymmetry.

2.2.1 Histogram

The *histogram* is a common way to represent the experimental distribution of the data. Consider a batch with n measurements that are sorted in increasing order, $z_1 < z_2 < \cdots < z_n$. The interval between the largest and the smallest value is divided into m bins (intervals) by the points $a_0, a_1, \ldots, a_{m-1}, a_m$. The intervals are usually of equal length and selected so that the histogram is relatively free from abrupt ups and downs. A measurement z belongs to the k-th bin if

$$a_{k-1} \leq z < a_k. \tag{2.1}$$

Define n_k as the number of measurements that belong to the k-th interval. The ratio n_k/n represents the frequency of occurrence in the k-th interval. Plotting the number n_k or the frequency of occurrences n_k/n as a bar diagram results in the histogram. See, for example, Figures 2.3, 2.4, and 2.5.

The histogram is probably the most widely recognized way to portray a data distribution. However, it has a potentially serious drawback: The visual impression conveyed may depend critically on the choice of the intervals. From Figure 2.3, we can see that the histogram is useless for small data sets, such as that of Table 1.1. For this reason, in many applications the box plot (which we will see later) is a better way to represent in a concise yet informative

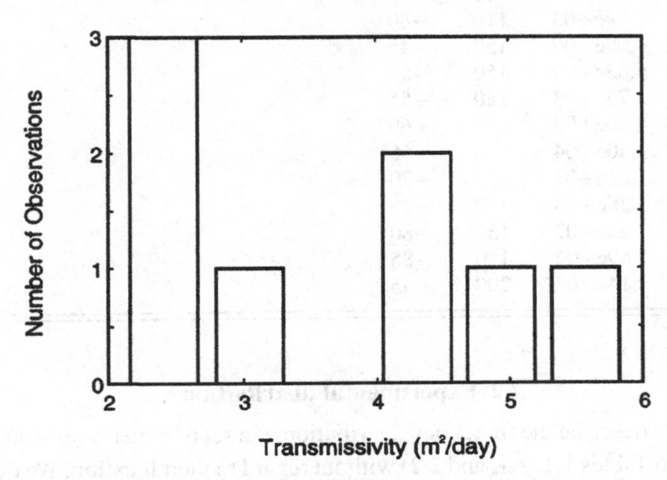

Figure 2.3 Histogram of transmissivity data.

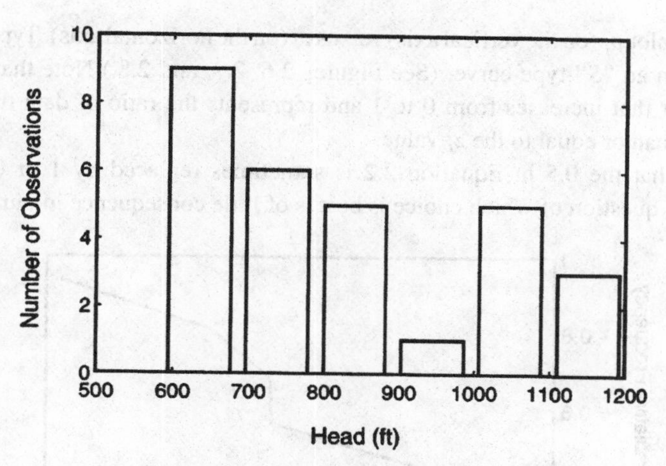

Figure 2.4 Histogram of head data.

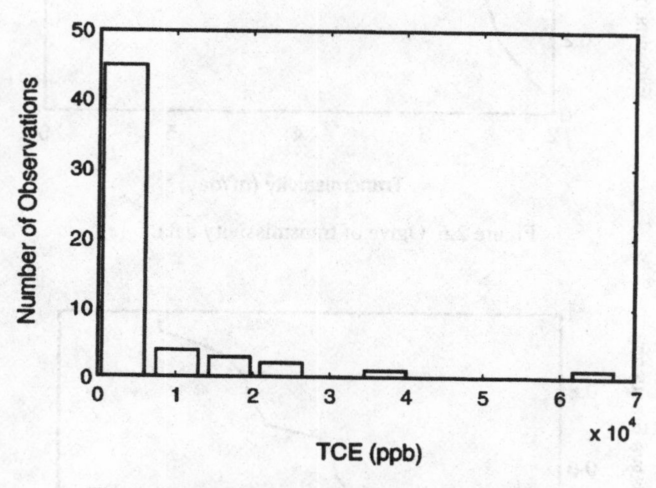

Figure 2.5 Histogram of TCE concentration data.

way the experimental distribution of the data, particularly if the number of measurements is small (*e.g.*, less than 50).

2.2.2 Ogive

The *ogive* is the experimental cumulative distribution. For the sorted data, $z_1 < z_2 < \cdots < z_n$, we compute

$$p_i = (i - 0.5)/n, \quad \text{for } i = 1, \ldots, n, \tag{2.2}$$

and then plot p_i (on the vertical axis) versus z_i (on the horizontal axis). Typically, we obtain an "S"-type curve. (See Figures 2.6, 2.7, and 2.8.) Note that p_i is a number that increases from 0 to 1 and represents the ratio of data that are smaller than or equal to the z_i value.

Note that the 0.5 in Equation 2.2 is sometimes replaced by 1 or 0. The technical question of which choice is best is of little consequence in our work.

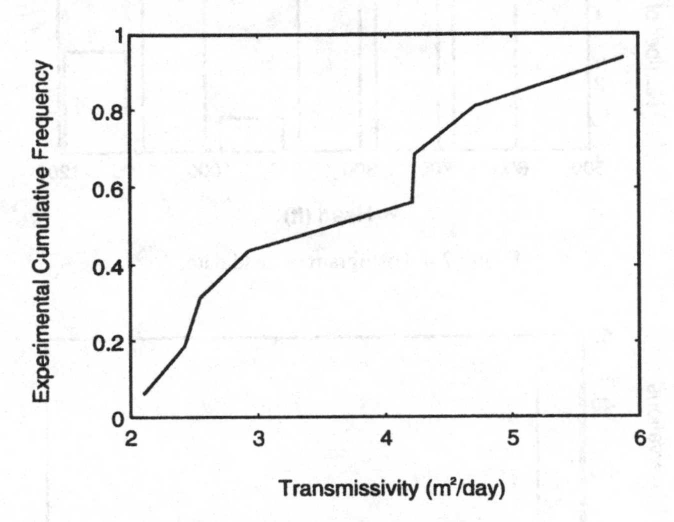

Figure 2.6 Ogive of transmissivity data.

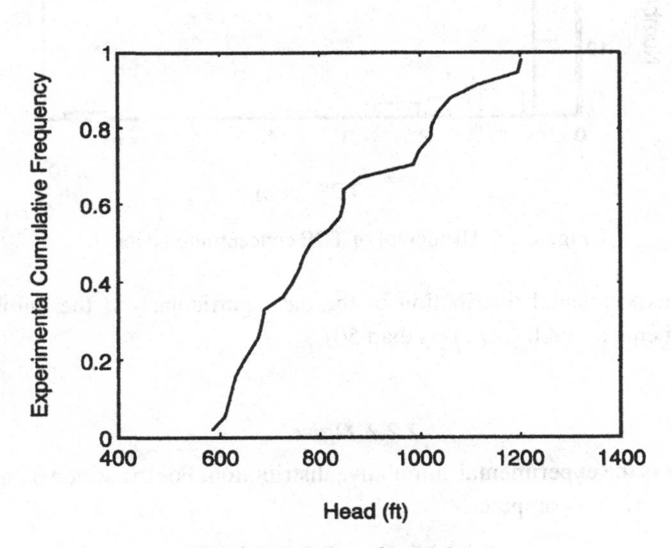

Figure 2.7 Ogive of head data.

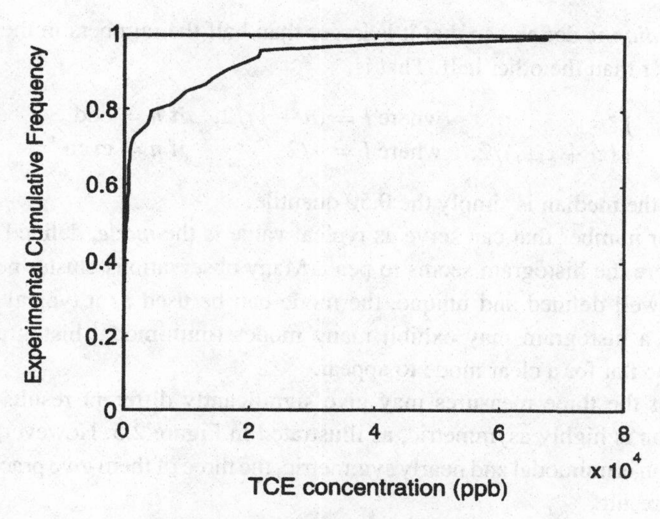

Figure 2.8 Ogive of concentration data.

A related concept is that of the quantile (a term that has a similar meaning to that of percentile). The p quantile (where p is a number between 0 and 1) is defined as the number $Q(p)$ that exceeds $100p$ percent of the data. For the discrete values $p_i = (i - 0.5)/n$, where $i = 1, \ldots, n$, the p_i quantile of the data is the value z_i. For other values of p, we interpolate linearly. For example, for p between p_i and p_{i+1},

$$Q(p) = Q(p_i + \nu(p_{i+1} - p_i)) = (1 - \nu)z_i + \nu\, z_{i+1}, \tag{2.3}$$

where ν is a number between 0 and 1, $\nu = (p - p_i)/(p_{i+1} - p_i)$.

2.2.3 Summary statistics

Usually, we are interested in a few numbers that summarize conveniently the most important features of the experimental distribution of the data. These are known as *summary statistics.*

2.2.3.1 Representative value

First, we are interested in a number that can be treated as a "typical" or "central" value, *i.e.*, a representative of the values in the batch. The most common such statistics are defined below.

The *arithmetic mean* or average of the batch is

$$\bar{z} = \frac{z_1 + z_2 + \cdots + z_n}{n} = \frac{1}{n} \sum_{i=2}^{n} z_i. \tag{2.4}$$

The *median* is defined so that it is larger than half the numbers in the batch and smaller than the other half. That is,

$$z_m = \begin{cases} z_l, & \text{where } l = (n+1)/2, \quad \text{if } n = \text{odd} \\ (z_l + z_{l+1})/2, & \text{where } l = n/2, \quad \text{if } n = \text{even} \end{cases}. \tag{2.5}$$

Note that the median is simply the 0.50 quantile.

Another number that can serve as typical value is the *mode*, defined as the value where the histogram seems to peak. Many observations cluster near the mode. If well defined and unique, the mode can be used as a typical value. However, a histogram may exhibit many modes (multimodal histogram) or may be too flat for a clear mode to appear.

Each of the three measures may give significantly different results if the distribution is highly asymmetric, as illustrated in Figure 2.5. However, if the distribution is unimodal and nearly symmetric, the three of them give practically the same result.

2.2.3.2 Spread

After the selecion of a representative value, one is interested in obtaining a measure of the spread of observations in the data set. A popular choice is the mean square difference from the arithmetic mean:

$$s^2 = \frac{(z_1 - \bar{z})^2 + \cdots + (z_n - \bar{z})^2}{n} = \frac{1}{n} \sum_{i=1}^{n} (z_i - \bar{z})^2. \tag{2.6}$$

s^2 is known as the batch *variance*; its square root, s, is the *standard deviation*.

Another measure of spread is the *interquartile range*, I_q (also known as the *Q-spread*). The interquartile range is simply

$$I_q = Q(0.75) - Q(0.25), \tag{2.7}$$

that is, the difference between the 0.75 and 0.25 quantiles. Note that $Q(0.75)$ is known as the *upper quartile* and $Q(0.25)$ as the *lower quartile*. An advantage of the interquartile range is that it is less sensitive to a few extreme values than the standard deviation. For this reason, the Q-spread is preferred in exploratory analysis whereas the standard deviation is used when the data follow an approximately normal distribution.

2.2.3.3 Symmetry

Of all the other characteristics of the distribution, symmetry or its absence is the most striking and important. For example, when the spread is characterized with a single number, it is implicitly assumed that the data are symmetrically distributed about the central value. Thus, it is useful to establish the degree of symmetry.

A useful measure in evaluating symmetry is the skewness coefficient

$$k_s = \left(\frac{1}{n} \sum_{i=1}^{n} (z_i - \bar{z})^3 \right) \Big/ s^3, \tag{2.8}$$

which is a dimensionless number. A symmetric distribution has k_s zero; if the data contain many values slightly smaller than the mean and a few values much larger than the mean (like the TCE concentration data), the coefficient of skewness is positive; if there are many values slightly larger and a few values much smaller than the mean, the coefficient of skewness is negative.

The statistics that summarize the important characteristics of the data are presented in Tables 2.3, 2.4, and 2.5.

Table 2.3. *Summary statistics for transmissivity data of Table 1.1*

Number of observations	8
Minimum value	2.1
First quartile	2.42
Median	3.57
Third quartile	4.24
Interquartile range	1.81
Maximum value	5.88
Mean	3.62
Standard deviation	1.33
Skewness coefficient	0.32

Table 2.4. *Summary statistics for head data of Table 2.1*

Number of observations	29
Minimum value	584
First quartile	662
Median	782
Third quartile	998
Interquartile range	336
Maximum value	1,202
Mean	830
Standard deviation	186
Skewness coefficient	0.49

Table 2.5. *Summary statistics for concentration data of Table 2.2*

Number of observations	56
Minimum value	2.5
First quartile	40
Median	435
Third quartile	2,540
Interquartile range	2,500
Maximum value	67,700
Mean	4,719
Standard deviation	11,300
Skewness coefficient	3.72

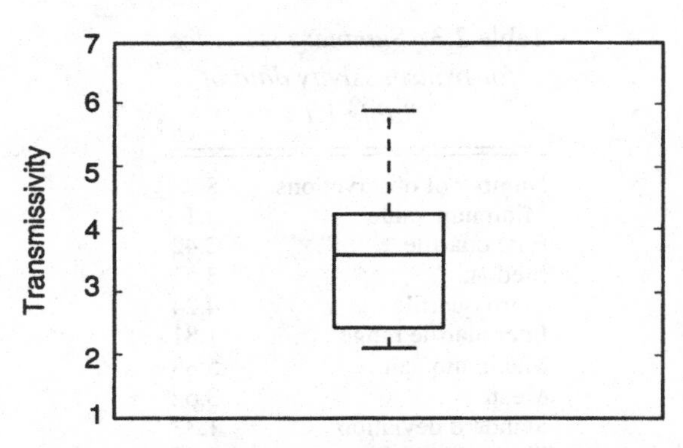

Figure 2.9 Box plot of transmissivity of data.

2.2.4 The box plot

The *box plot* (see Figures 2.9, 2.10, and 2.11) is a visually effective way to summarize graphically the distribution of the data. The upper and lower quartiles of the data define the top and bottom of a rectangle (the "box"), and the median is portrayed by a horizontal line segment inside the box. From the upper and lower sides of the rectangle, dashed lines extend to the so-called adjacent values or fences. The upper adjacent value is the largest observed value provided that the length of the dashed line is smaller than 1.5 times the interquartile range; otherwise, we just draw the dashed lines 1.5 times the interquartile range and we plot all observations that exceed the upper adjacent point as little circles or asterisks. Exactly the same procedure is followed for the lower adjacent value. Observations outside of the range between the adjacent values are known as *outside values*.

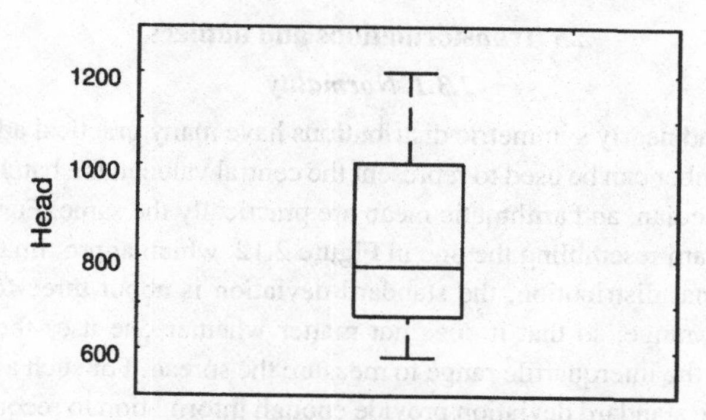

Figure 2.10 Box plot of head data.

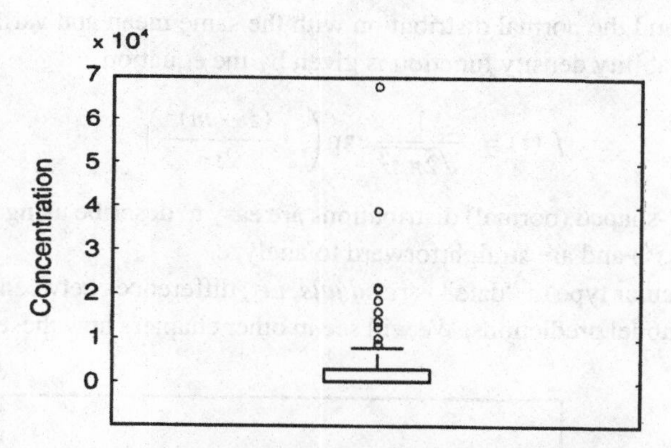

Figure 2.11 Box plot of TCE concentration data.

You must realize that there is nothing magical about the number $1.5 \times I_q$ used in the criterion for outside values. It is a useful convention (one of many in statistics) and is to some extent arbitrary. The essence of the criterion is that for normally distributed data, the probability of a measurement being outside of the thus defined fences is very small.

The box plot is a graph of the key characteristics of the data distribution. The line inside the box (location of the median) represents the center of the batch. The size of the box represents the spread about the central value. One may judge whether the data are distributed symmetrically by checking whether the median is centrally located within the box and the dashed lines are of approximately the same length. The lengths of the dashed lines show how stretched the tails of the histogram are. Finally, the circles or asterisks indicate the presence of outside values.

2.3 Transformations and outliers

2.3.1 Normality

Unimodal and nearly symmetric distributions have many practical advantages. A single number can be used to represent the central value in the batch, because the mode, median, and arithmetic mean are practically the same. Furthermore, for a histogram resembling the one in Figure 2.12, which approximates a bell-shaped normal distribution, the standard deviation is about three-fourths the interquartile range, so that it does not matter whether one uses the standard deviation or the interquartile range to measure the spread. For such a batch, the mean and the standard deviation provide enough information to reconstruct the histogram with acceptable accuracy.

Figure 2.12 shows the histogram of some hypothetical data with mean m and variance s^2 and the normal distribution with the same mean and variance. The normal probability density function is given by the equation

$$f(z) = \frac{1}{\sqrt{2\pi s^2}} \exp\left(-\frac{(z-m)^2}{2s^2}\right). \tag{2.9}$$

Thus, bell-shaped (normal) distributions are easy to describe using two numbers (m and s^2) and are straightforward to analyze.

One particular type of "data" is *residuals*, *i.e.*, differences between observed values and model predictions. We will see in other chapters how these residuals

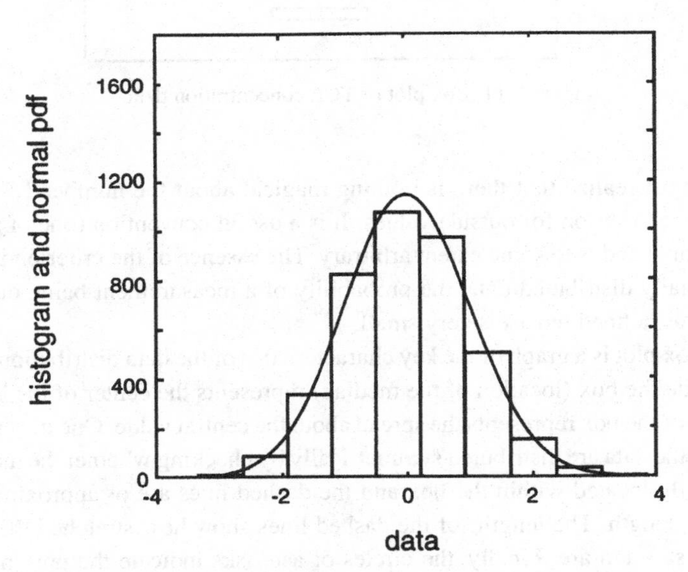

Figure 2.12 Histogram and theoretical distribution of normal data.

are computed and why they are important. For now, we mention that the commonly used linear estimation methods make most sense when the distribution of the residuals is approximately normal. For a normal distribution, the mean is the indisputable central value and the standard deviation is the indisputable measure of spread. That is why setting the mean of the residuals to zero and minimizing the standard deviation of the errors (which is what linear estimation methods do) is clearly a good way to make predictions.

As illustrated by the concentration observations of Table 2.2, data distributions may not be unimodal and symmetric. Given the advantages of nearly normal batches, it is reasonable to search for a simple transformation that makes it possible to describe a distribution with a mean and a variance. For example, for concentration and transmissivity data that are positive, one may use the so-called *power transformation*:

$$y = \begin{cases} (z^\kappa - 1)/\kappa & \kappa > 0 \\ \ln(z) & \end{cases} . \tag{2.10}$$

Note that the commonly used logarithmic transformation is included as a special case of the power transformation, obtained at the limit for $\kappa = 0$. An application of the logarithm transformation is found in reference [134].

For example, Figure 2.13 shows the box plot of the logarithm of the concentration data. The distribution of the transformed data is much easier to describe than the distribution of the original data. Thus, one can summarize the important characteristics of a data set through the parameter κ and the mean and variance of the transformed data.

An important point remains to be made. The basic assumption in the type of methods that we will see later is that the *estimation errors* are approximately normally distributed. Thus, the purpose of the transformation is to adjust

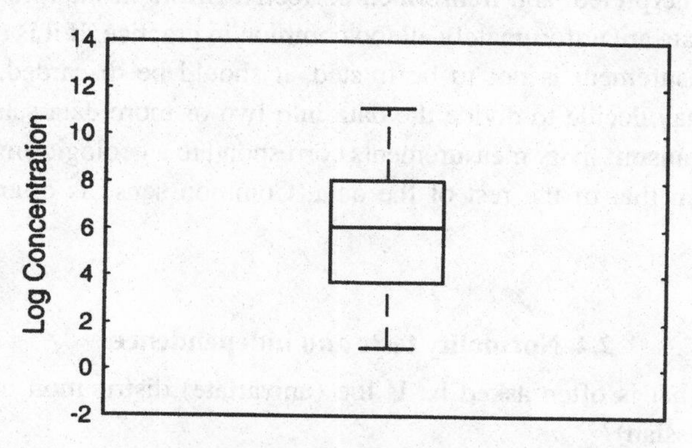

Figure 2.13 Box plot of the logarithm of the concentration data.

the distribution of some estimation errors (residuals) and not necessarily the distribution of the data. Later, we will test the box plot residuals to ascertain whether they are nearly normally distributed.

2.3.2 *Outliers*

Let us now discuss the often misunderstood issue of outside values; commonly referred to as *outliers*, they are surrounded by an aura of mystique. The box plot of the concentration measurements (Figure 2.11) calls attention to values that seem to stand out from the bulk of data. All this means, however, is that it would be unwise at this preliminary stage to lump those "unusual" values with the other data and then to compute their mean and standard deviation. The mean and the variance would be excessively affected by the relatively few outside values and would not be representative of either the bulk of the data or the outside values. However, it would be equally unwise to brand uncritically these unusual measurements as "outliers," "stragglers," or "stray" and to discard them.

If the box plot indicates that the data are distributed in a highly asymmetric way or that the tails are stretched out, it is possible that after a transformation all or some of the values that were originally outside will get inside. In this case, the practical issue is to find a good way to describe the asymmetric distribution of data, such as by transforming to a more symmetric distribution. In many cases, including the concentration data of Table 2.2, the so-called outliers are the most interesting measurements in the data set.

Before deciding what to do with an outside value, one must return to the source of the data and use one's understanding of the physical processes involved. A reasonable effort should be made to verify that the measurement was taken, interpreted, and transcribed correctly. Errors in interpretation and copying of data are unfortunately all too common in practice. If it is concluded that the measurement is not to be trusted, it should be discarded. In some cases, one may decide to divide the data into two or more data sets, such as when stray transmissivity measurements correspond to a geologic environment different from that of the rest of the data. Common sense is often the best guide.

2.4 Normality tests and independence

A question that is often asked is: Is the (univariate) distribution of the data normal (Gaussian)?

One approach involves applying *statistical hypothesis tests* (some of which will be discussed elsewhere). There are general tests that can be applied to any distribution, such as the *chi-square* and the *Kolmogorov-Smirnov* tests, and there are tests that have been developed specifically for the normal distribution, such as variations of the *Shapiro-Wilks* test [126]. Each of these tests provides a procedure to determine whether the data depart sufficiently from the *null hypothesis* that the observations were sampled independently from a normal distribution.

The key limitation of these tests in exploratory analysis is that they assume that the data were independently generated, *i.e.*, that there is negligible correlation among observations. However, as will be seen later, correlation is usually important, such as when two measurements tend to differ by less as the distance between their locations decreases. In this case, a statistical test that assumes independence is not appropriate.

Actually, these tests should be applied to *orthonormal residuals*, which are differences between observations and predictions. These residuals are supposed to be uncorrelated and to follow a normal distribution with zero mean and unit variance. The next two chapters describe how to compute these residuals and how to apply such normality tests.

2.5 Spatial structure

The goal of the data analysis methods of Sections 2.2–2.4 was to describe the distribution of measurements independently of their location. What about describing how measurements vary in space or are related based on their location? For example: Are neighboring measurements more likely to be similar in value than distant ones? Do observed values increase in a certain direction?

Data, such as those of Tables 1.1, 1.2, and 2.2, should be displayed in ways that reveal their spatial structure. First, a word of caution is warranted. Nowadays, there are commercial software packages that can draw contour maps or three-dimensional mesh surfaces of a variable $z(x_1, x_2)$ directly from the observations and even plot a variable $z(x_1, x_2, x_3)$ that varies in three dimensions. As an intermediate step, these packages interpolate from the measurements to a fine mesh, which is needed for contouring. If the observations are few, are nonuniformly distributed, and have highly skewed distributions, then the plot obtained is affected by the interpolation and smoothing routines used in the program, not just by the data. Important features may be smoothed out; puzzling results may appear in areas with few measurements as artifacts of the algorithms used. The software user usually has little control over the interpolation algorithms and sometimes is not even informed what methods are used. One may still find it useful to apply such packages for exploratory data visualization but not to produce final results.

If a variable depends on only one spatial dimension (such as measurements of chemical concentrations in a boring or flowmeter measurements of conductivity in a borehole) then the most powerful plot is obviously that of observation against location, $z(x_i)$ against x_i. Even for variables that depend on two or three spatial dimensions, one may start with x-y plots of the observations against each spatial coordinate. For example, see Figures 2.14 and 2.15 for the head data.

For two-dimensional variability, $z(x_1, x_2)$, a particularly useful way to visualize the data is by plotting an (x_1, x_2) scatter plot showing the location of each measurement with symbols indicating whether the measured value is above or below the median, whether it is a stray value, and other pertinent information. For example, such a plot in the case of the head data, Figure 2.16, suggests an apparent trend.

For three-dimensional variability, one can use the *draftsman's display*, which consists of three pairwise scatter plots (x_1, x_2), (x_1, x_3), and (x_2, x_3), arranged so that adjacent plots share an axis. The idea is that the three plots represent views from the top, the front, and the side of points in a rectangular parallelepiped that contains the measurement locations. With careful choice of symbols and considerable effort, one may see some patterns. For example, for the data plotted in Figure 2.17, the draftsman display shows that the measured values tend to increase in value in the direction of increasing x_1 and x_2.

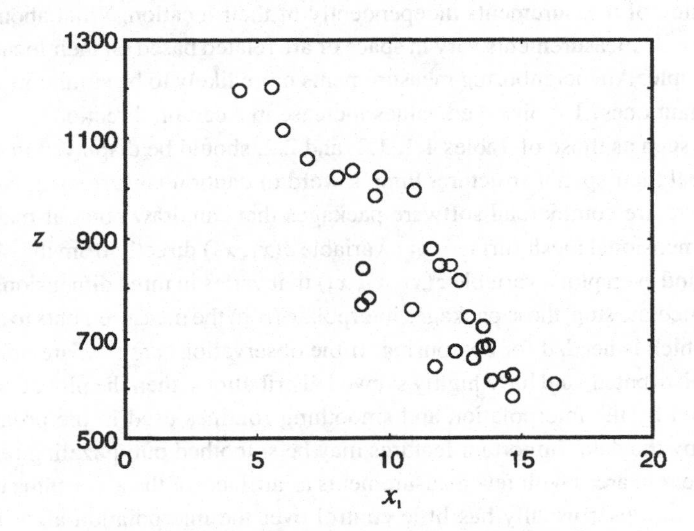

Figure 2.14 Plot of data versus the first spatial coordinate.

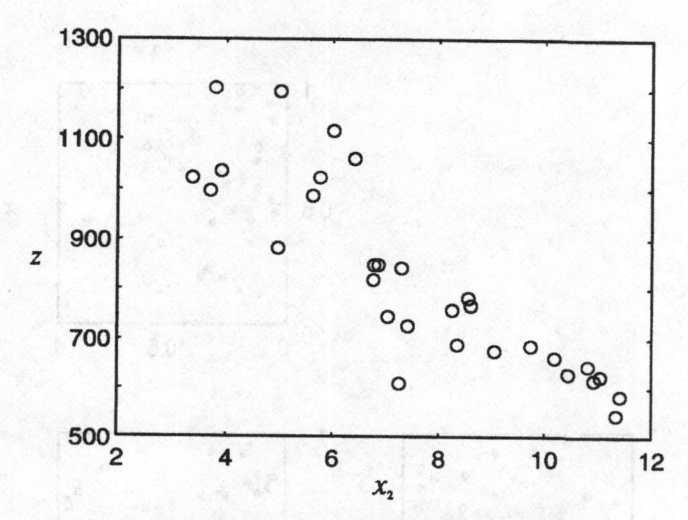

Figure 2.15 Plot of data versus the second spatial coordinate.

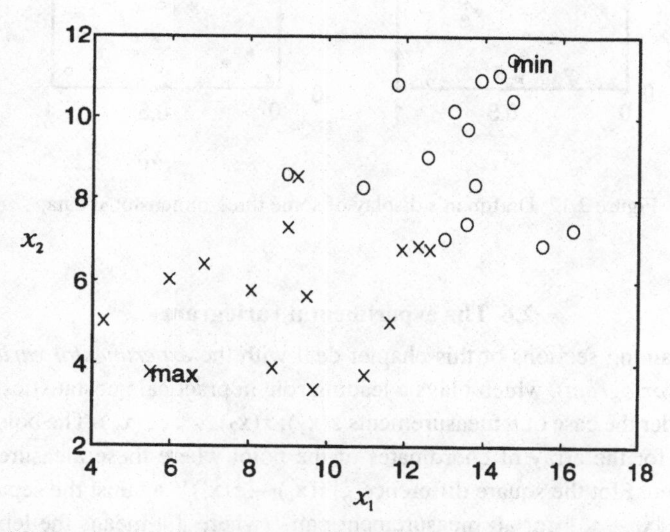

Figure 2.16 Plot showing the location and relative magnitude of data (○ < median, × > median). When outliers are present, they are indicated by *.

Three-dimensional graphics are gradually becoming available with perspective, movement, and shading for better visualization. Data are represented as spheres or clouds of variable size or color indicative of magnitude of the observation.

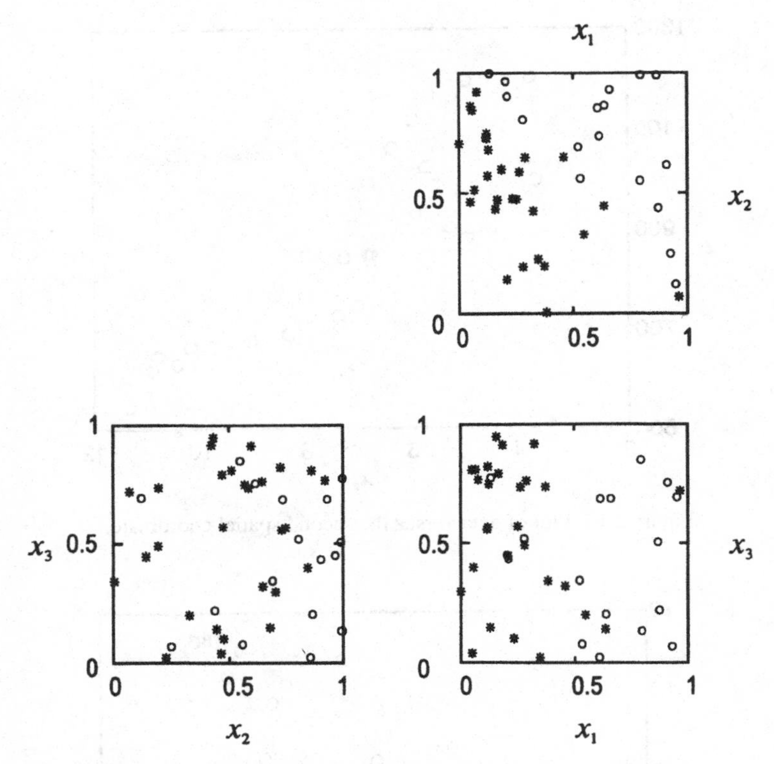

Figure 2.17 Draftman's display of some three-dimensional data.

2.6 The experimental variogram

The remaining sections of this chapter deal with the *experimental variogram* (or *semivariogram*), which plays a leading role in practical geostatistics.

Consider the case of n measurements $z(\mathbf{x}_1), z(\mathbf{x}_2), \ldots, z(\mathbf{x}_n)$. The bold letter $\mathbf{x}$ stands for the array of coordinates of the point where these measurements were taken. Plot the square difference $\frac{1}{2}[z(\mathbf{x}_i) - z(\mathbf{x}_i')]^2$ against the separation distance $\|\mathbf{x}_i - \mathbf{x}_i'\|$ for all measurement pairs (where $\| \; \|$ means the length of a vector). For n measurements, there are $\frac{n(n-1)}{2}$ such pairs that form a scatter plot known as the *raw variogram* (represented by the dots in Figure 2.18 for the head data). The experimental variogram is a smooth line through this scatter plot.

In the common method of plotting the experimental variogram, the axis of separation distance is divided into consecutive intervals, similarly as for the histogram. The k-th interval is $[h_k^\ell, h_k^u]$ and contains N_k pairs of measurements

$[z(\mathbf{x}_i), z(\mathbf{x}'_i)]$. Then, compute

$$\hat{\gamma}(h_k) = \frac{1}{2N_k} \sum_{i=1}^{N_k} [z(\mathbf{x}_i) - z(\mathbf{x}'_i)]^2, \tag{2.11}$$

where index i refers to each pair of measurements $z(\mathbf{x}_i)$ and $z(\mathbf{x}'_i)$ for which

$$h_k^\ell \leq \|\mathbf{x}_i - \mathbf{x}'_i\| < h_k^u. \tag{2.12}$$

This interval is represented by a single point h_k. Take h_k equal to the average value,

$$h_k = \frac{1}{N_k} \sum_{i=1}^{N_k} \|\mathbf{x}_i - \mathbf{x}'_i\|. \tag{2.13}$$

Next, these points $[h_k, \hat{\gamma}(h_k)]$ are connected to form the experimental variogram (see Figures 2.18 and 2.19). Modifications to this basic approach have been proposed to improve its robustness [5, 31, 30, 37, 107].

In selecting the length of an interval, keep in mind that by increasing the length of the interval you average over more points, thus decreasing the fluctuations of the raw variogram, but you may smooth out the curvature of the variogram. It is unprofitable to spend too much time at this stage fiddling with the intervals because there is really no "best" experimental variogram. Some useful guidelines to obtain a reasonable experimental variogram are:

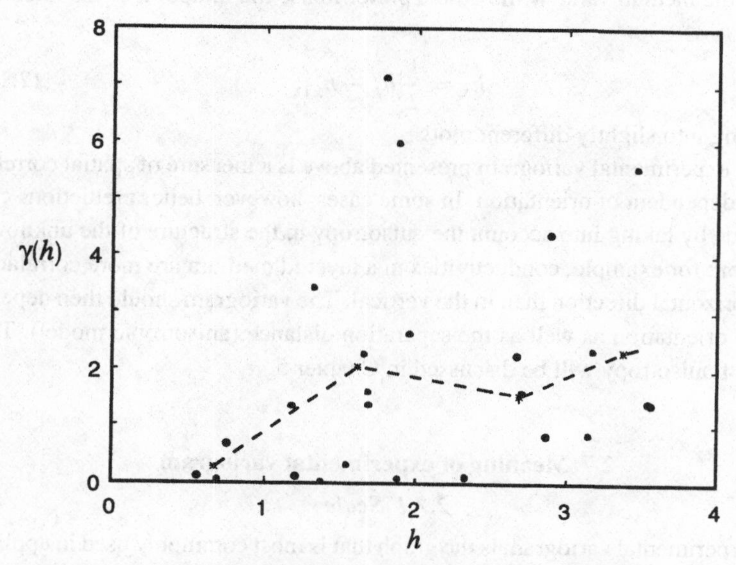

Figure 2.18 Raw and experimental variogram of transmissivity data.

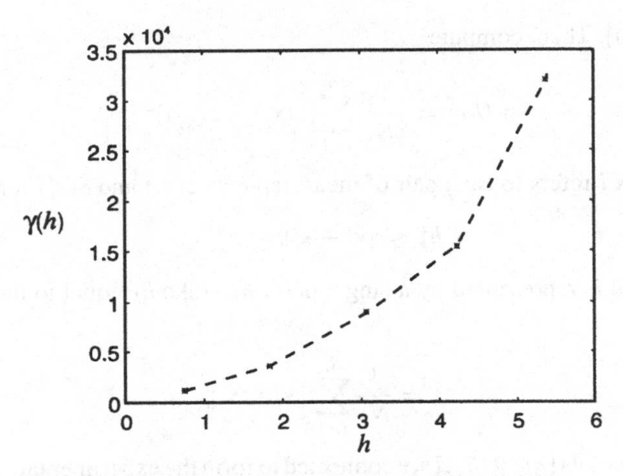

Figure 2.19 Experimental variogram of the head data.

1. Use three to six intervals.
2. Include more pairs (use longer intervals) at distances where the raw variogram is spread out.

As an exploratory analysis tool, the experimental variogram has the drawback that the graph depends on the selected intervals. It may also be somewhat affected by the method of averaging. For example, some analysts prefer to use for h_k the median value while others prefer to use the midpoint of the interval, *i.e.*,

$$h_k = \frac{1}{2}\left|h_k^u - h_k^\ell\right|, \tag{2.14}$$

resulting into slightly different plots.

The experimental variogram presented above is a measure of spatial correlation independent of orientation. In some cases, however, better predictions can be made by taking into account the anisotropy in the structure of the unknown function; for example, conductivities in a layered medium are more correlated in a horizontal direction than in the vertical. The variogram should then depend on the orientation as well as the separation distance (anisotropic model). The issue of anisotropy will be discussed in Chapter 5.

2.7 Meaning of experimental variogram

2.7.1 Scale

The experimental variogram is the graph that is most commonly used in applied geostatistics to explore spatial interdependence. It contains information about the *scale* of fluctuations of the variable, as we will explain in this section. Some readers may prefer to skip this section at first reading and come back to it after Chapter 3 or 4.

To grasp the concept of scale, consider the function z that varies over a one-dimensional domain x. It is useful to approximate $z(x)$ by the sum of cosines

$$z(x) \simeq \sum_i A_i \cos(2\pi x / L_i + \phi_i), \tag{2.15}$$

where A_i is coefficient of the trigonometric series, L_i is spatial scale, and ϕ_i is phase shift. Without getting too technical and disregarding pathological cases, practically any function can be sufficiently approximated over a finite domain by a few terms of a trigonometric, or Fourier, series. If the length of the domain is L, then $1/L_i = i/L$, where $i = 0, 1, 2, \ldots$. When A_i^2 is relatively large, we say that a proportionally large part of the variability is "at scale L_i." One can reconstruct $z(x)$ from knowledge of the triplets (L_i, A_i, ϕ_i) for all terms in the series.

The bottom line is that the experimental variogram contains information about the A_i^2 values. That is, one can infer the approximate value of some of the A_i^2 values. Thus, the experimental variogram can provide clues on whether the scale of the variability is large or small (or what is known as the "power spectrum of the function").

However, the experimental variogram provides no real information about the phase shifts, *i.e.*, the variogram is not helpful in reconstructing the starting points of the cosinusoidal waves comprising the actual function. Two functions z that have the same variogram may look radically different because of different phase shifts. Also, the computation of the variogram scrambles or masks patterns in the data, such as clear trends, which might be easy to recognize from other plots. We have not attempted to prove these assertions mathematically because that would involve knowledge that cannot be assumed at this point. In other chapters, however, we will obtain insights that will support the statement that the experimental variogram is basically a way to infer the distribution of spatial variability with respect to spatial scales.

We need to emphasize that although it is an important exploratory analysis tool, the experimental variogram should not monopolize the analysis. It should be used in conjunction with other analysis tools, such as those presented in Section 2.4.

Data analysts are particularly interested in two structural characteristics:

- The presence of variability at the scale of the sampling span. This depends on the behavior of the experimental variogram near the origin, *i.e.*, at small separation distances.

- The presence of variability at a scale comparable to the sampling domain. This depends on the behavior of the experimental variogram at large distances.

2.7.2 Near the origin

The behavior of the variogram at small separation distances determines whether the spatial function appears continuous and smooth. We will consider three examples, which are intended to give you an intuitive feeling about what we mean by continuity and smoothness in a practical context.

2.7.2.1 Discontinuous

First, consider the case that the actual variable is

$$z_1(x) = \cos(2\pi x/0.001). \tag{2.16}$$

That is, all the variability is at scale 0.001. Consider now that $z_1(x)$ is sampled at 100 locations randomly distributed in the interval between 0 and 1. (The same sampling locations will be used in all three examples.) Note that the average sampling interval (*i.e.*, distance between measurement locations), 0.01, is ten times larger than the scale of fluctuations of the variable. As a result, two adjacent measurements are about as different as two distant measurements. At the scale of the sampling interval, the variable z_1 is discontinuous because it changes abruptly from one sampling point to the next, as shown in Figure 2.20.

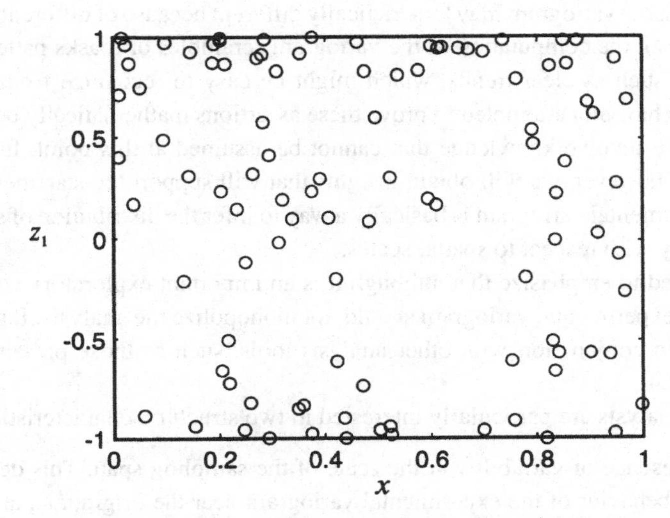

Figure 2.20 Plot of the measurements for the discontinuous case.

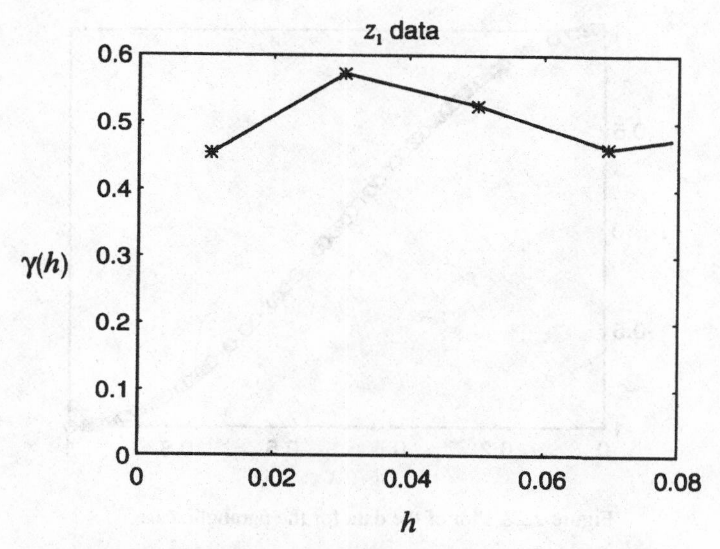

Figure 2.21 Experimental variogram for the discontinuous data.

The experimental variogram, shown in Figure 2.21, is approximately a straight horizontal line. Because the experimental variogram does not seem to converge to zero as the separation decreases, we say that there is a *discontinuity of the experimental variogram at the origin* or a *nugget effect*.

In general, a discontinuity at the origin in the experimental variogram is indicative of fluctuations at a scale smaller than the sampling interval, called *microvariability*. It may also be due to *random observation error*, as we will discuss further elsewhere.

2.7.2.2 Parabolic

As a second example, consider

$$z_2(x) = \cos(2\pi x/2). \tag{2.17}$$

All the variability is at a scale much larger than the scale of the sampling intervals. Figure 2.22 is a plot of the data $z_2(x_m)$ versus x_m and Figure 2.23 is a plot of the slopes of the data $\frac{z_2(x_m)-z_2(x_{m-1})}{x_m-x_{m-1}}$ versus x_{m-1}, for all $m = 2, \ldots, 100$. The changes in measured values are so gradual that both z and its slope are observed to vary continuously. The experimental variogram, shown on Figure 2.24, has parabolic behavior near the origin; that is, it is proportional to h^2 for small values of h. Generally, parabolic behavior near the origin is indicative of a quantity that is smooth at the scale of the measurements so that it is differentiable (*i.e.*, it has a well-defined slope).

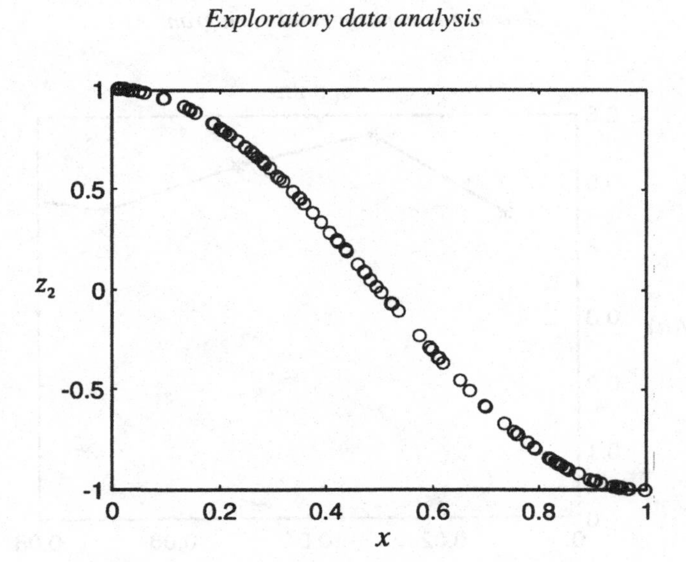

Figure 2.22 Plot of the data for the parabolic case.

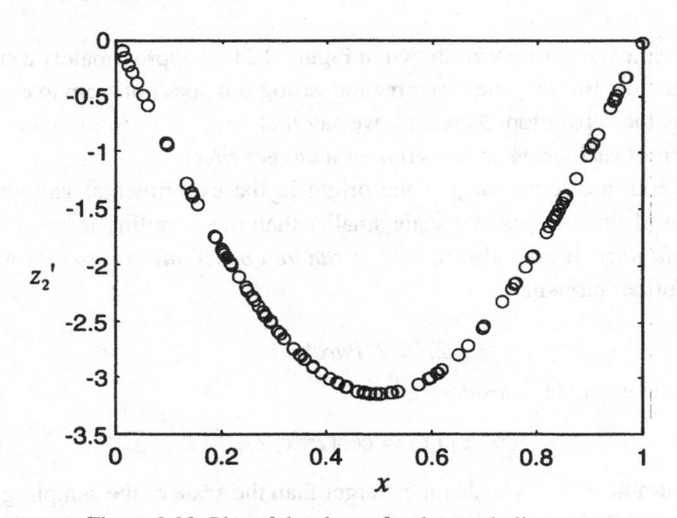

Figure 2.23 Plot of the slopes for the parabolic case.

2.7.2.3 Linear

As a third example, we will consider a case in between the first two examples. Consider that

$$z_3(x) = \cos(2\pi x/0.2) + 0.10\cos(2\pi x/0.02) \qquad (2.18)$$

with the same sampling points. Note that this variable has most of its variability at a scale larger than the average sampling interval but also some variability at a scale comparable to that of the measurement spacing. The changes in the value of z_3 between adjacent sampling points are gradual, as shown on Figure 2.25, so z_3 is practically continuous at the scale of the measurements.

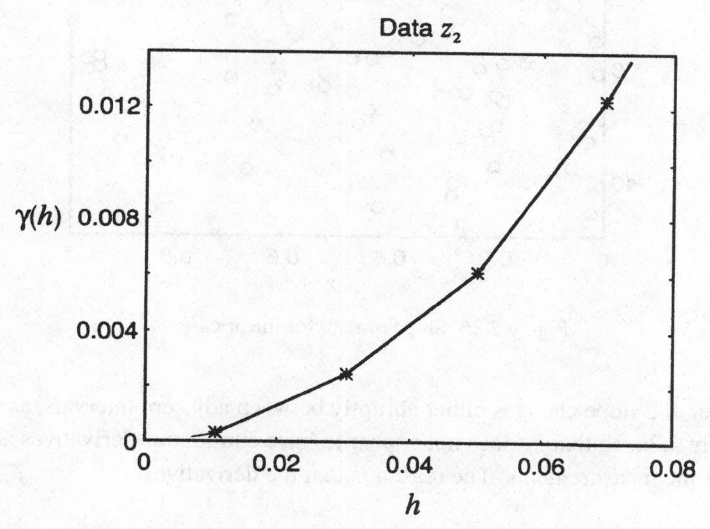

Figure 2.24 Experimental variogram for the parabolic case.

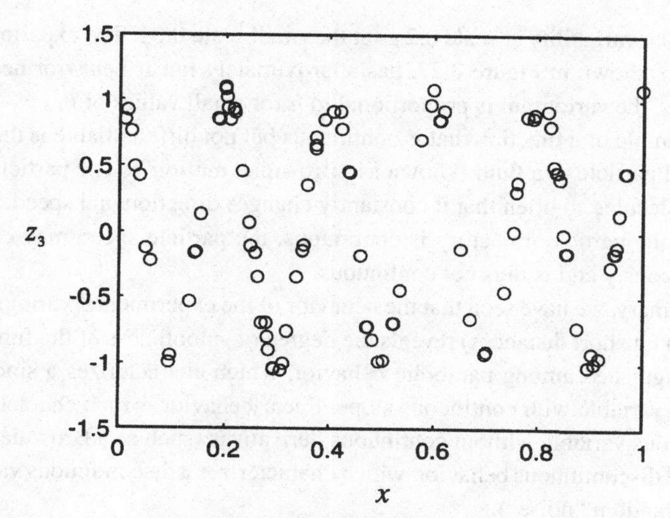

Figure 2.25 The data for the linear case.

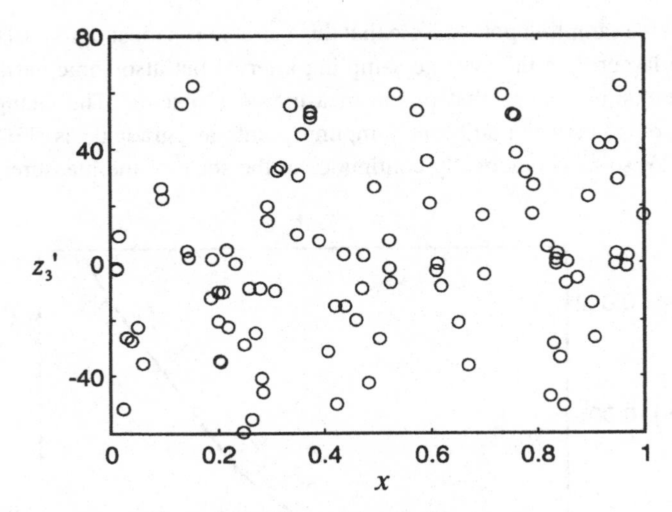

Figure 2.26 Slope of data for linear case.

However, the slope changes rather abruptly between adjacent intervals, as seen in Figure 2.26, so that z_3 does not appear to have continuous derivatives at the scale of the measurements. The reason is that the derivative

$$\frac{dz_3}{dx} = -10\pi \sin(2\pi x/0.2) - 10\pi \sin(2\pi x/0.02) \tag{2.19}$$

has as much variability at scale 0.2 as at the small scale 0.02. The experimental variogram, shown in Figure 2.27, has approximately linear behavior near the origin, *i.e.*, the variogram is proportional to h for small values of h.

An example of a function that is continuous but not differentiable is the path of a small particle in a fluid (known as "Brownian motion"). The particle gets hit by molecules so often that it constantly changes direction and speed. Thus, although the particle trajectory is continuous, the particle speed may change instantaneously and is thus not continuous.

In summary, we have seen that the behavior of the experimental variogram at the origin (at short distances) reveals the degree of smoothness of the function. We distinguished among parabolic behavior, which characterizes a smoothly changing variable with continuous slope; linear behavior, which characterizes a continuous variable without continuous derivatives (such as a Brownian motion); and discontinuous behavior, which characterizes a discontinuous variable (such as random "noise").

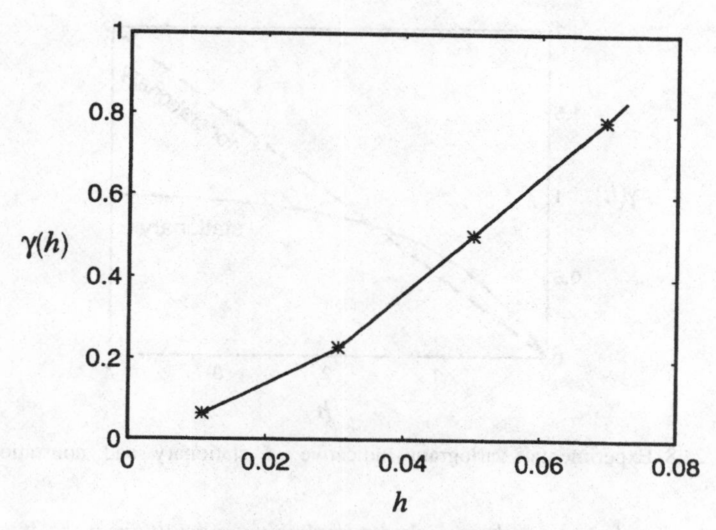

Figure 2.27 Experimental variogram for linear case.

2.7.3 Large-scale behavior

The behavior of the variogram at distances comparable to the size of the domain determines whether the function is *stationary*. We will later give a technical meaning to the term stationary; intuitively, a function is stationary if it consists of small-scale fluctuations (compared to the size of the domain) about some well-defined mean value. For such a function, the experimental variogram should stabilize around a value, called the *sill*, as shown in Figure 2.28. For a stationary function, the length scale at which the sill is obtained describes the scale at which two measurements of the variable become practically uncorrelated. This length scale is known as *range* or *correlation length*.

Otherwise, the variogram keeps on increasing even at a distance comparable to the maximum separation distance of interest, as shown in Figure 2.28.

Exercise 2.1 *Consider two functions describing a quantity that varies along the spatial coordinate x in the interval* [0, 4]:

$$z_1(x) = \cos\left(\frac{\pi x}{2}\right) + \frac{1}{3^2}\cos\left(\frac{3\pi x}{2}\right) + \frac{1}{5^2}\cos\left(\frac{5\pi x}{2}\right) + \cdots$$

$$z_2(x) = \cos\left(\frac{\pi x}{2} + \phi_1\right) + \frac{1}{3^2}\cos\left(\frac{3\pi x}{2} + \phi_3\right) + \frac{1}{5^2}\cos\left(\frac{5\pi x}{2} + \phi_5\right) + \cdots,$$

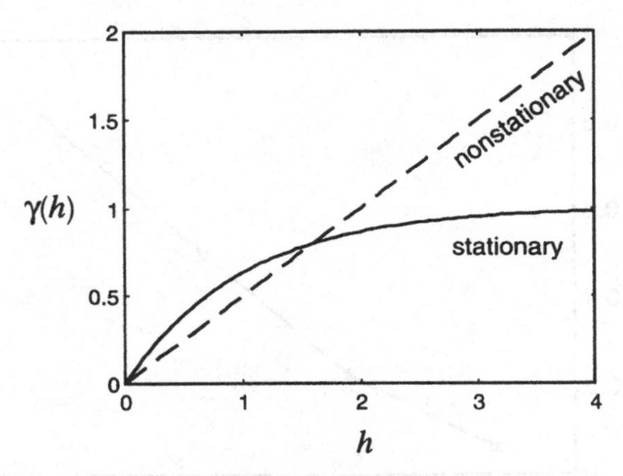

Figure 2.28 Experimental variogram indicative of stationary and nonstationary behavior.

where ϕ_1, ϕ_3, ϕ_5 *are numbers picked completely randomly from the interval* $[0, 2\pi]$. *Plot these two functions and discuss, based on what you have read in this chapter, how the experimental variograms of these two functions are expected to differ. Based on this example, discuss the strengths and limitations of the experimental variogram as an exploratory analysis tool.*

2.8 Key points of Chapter 2

The objective of exploratory analysis is to familiarize the analyst with the important characteristics of the data. The analyst should keep an open mind and avoid techniques that may be misleading if certain assumptions are not met. We start by analyzing the distribution of data independently of their location in space; this distribution may be portrayed using the histogram, the ogive, and the box plot. Important summary statistics are the median and the mean, the interquartile range and the standard deviation, and the skewness coefficient. We discussed the practical advantages of working with symmetric and nearly normal distributions and how transformations can be used to achieve this goal. Spatial variability can be analyzed using graphical techniques, but the difficulty increases significantly from variability in one dimension to variability in three dimensions. The experimental variogram is an important tool that provides information about the distribution of spatial variability with respect to scales. Finally, note that conclusions reached during an exploratory analysis are usually tentative. The next step is to use the ideas created during exploratory analysis to select tentatively an "equation to fit to the data."

3

Intrinsic model

We preview the general methodology underlying geostatistical modeling and apply it to the most common model, which is known as the *intrinsic isotropic model* and is characterized by the variogram. This chapter introduces *kriging*, which is a method for evaluating estimates and mean square estimation errors from the data, for a given variogram. The discussion in this chapter is limited to isotropic correlation structures (same correlation in all directions) and focuses on the methodology and the basic mathematical tools. Variogram selection and fitting will be discussed in the next chapter.

3.1 Methodology overview

Consider that we have measured porosity along a borehole at several locations (see Figure 3.1). To estimate the value of the porosity at any location from the measured porosity values, we need a mathematical expression (or "equation" or "model") that describes how the porosity varies with depth in the borehole. In other words, we need a model of spatial variability.

However, hydrologic and environmental variables change from location to location in complex and inadequately understood ways. In most applications, we have to rely on the data to guide us in developing an empirical model. The model involves the concept of probability in the sense that spatial variability is described coarsely by using averages. For example, the best we can do might be to specify that the porosity fluctuates about some mean value and to come up with a formula to correlate the fluctuations at two locations depending on their separation distance. This is often the most practical scheme to summarize incomplete information or erratic data.

Consider the porosity or any other spatially variable quantity, such as chemical concentration or precipitation; this quantity is a function of the spatial coordinates and may be represented as $z(x_1)$, $z(x_1, x_2)$, or $z(x_1, x_2, x_3)$ depending

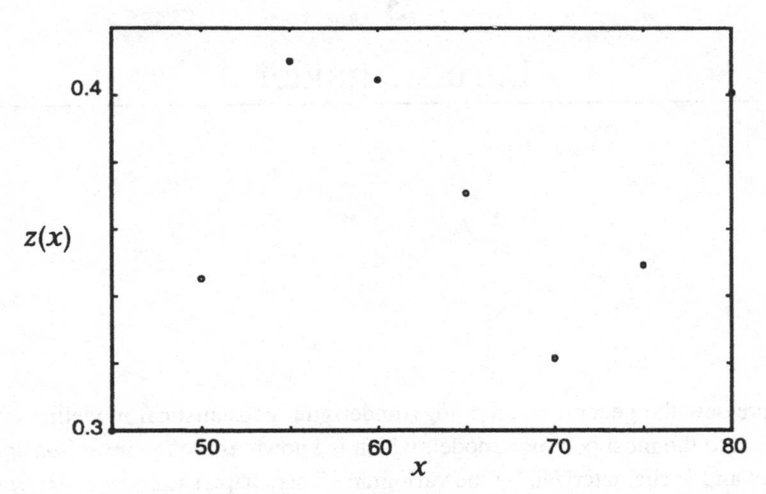

Figure 3.1 The interpolation problem. Observations are indicated by the symbol o.

on whether it varies in one, two, or three dimensions. For brevity, the notation $z(\mathbf{x})$ will be used to include all three cases, where $\mathbf{x}$ is the *location index* (a vector with one, two, or three components). Thus,

$$\mathbf{x} = x_1, \quad \text{or} \quad \mathbf{x} = \begin{bmatrix} x_1 \\ x_2 \end{bmatrix}, \quad \text{or} \quad \mathbf{x} = \begin{bmatrix} x_1 \\ x_2 \\ x_3 \end{bmatrix}. \tag{3.1}$$

The function $z(\mathbf{x})$, known as a *regionalized* or *field* variable, is not known everywhere but needs to be estimated from available observations and, perhaps, additional information.

We are now ready to discuss the logical underpinnings of the approach. If statistical modeling is new to you and you wonder what it means, pay particular attention to this part.

In practice, our objective is to estimate a field variable $z(\mathbf{x})$ over a region. Usually, because of scarcity of information, we cannot find a unique solution. It is useful to think of the actual unknown $z(\mathbf{x})$ as one out of a collection (or *ensemble*) of possibilities $z(\mathbf{x}; 1)$, $z(\mathbf{x}; 2)$, ... This ensemble defines all possible solutions to our estimation problem. The members of the ensemble are known as *realizations* or *sample functions*.

Consider, for example, Figures 3.2, 3.3, and 3.4. Each figure contains five realizations from a different ensemble (family of functions). Notice that despite the differences among the realizations in each figure, they share some general structural characteristics. The functions in Figure 3.2 are all "smooth" curves with well-defined slope at every point. The functions in Figures 3.3 and 3.4 are

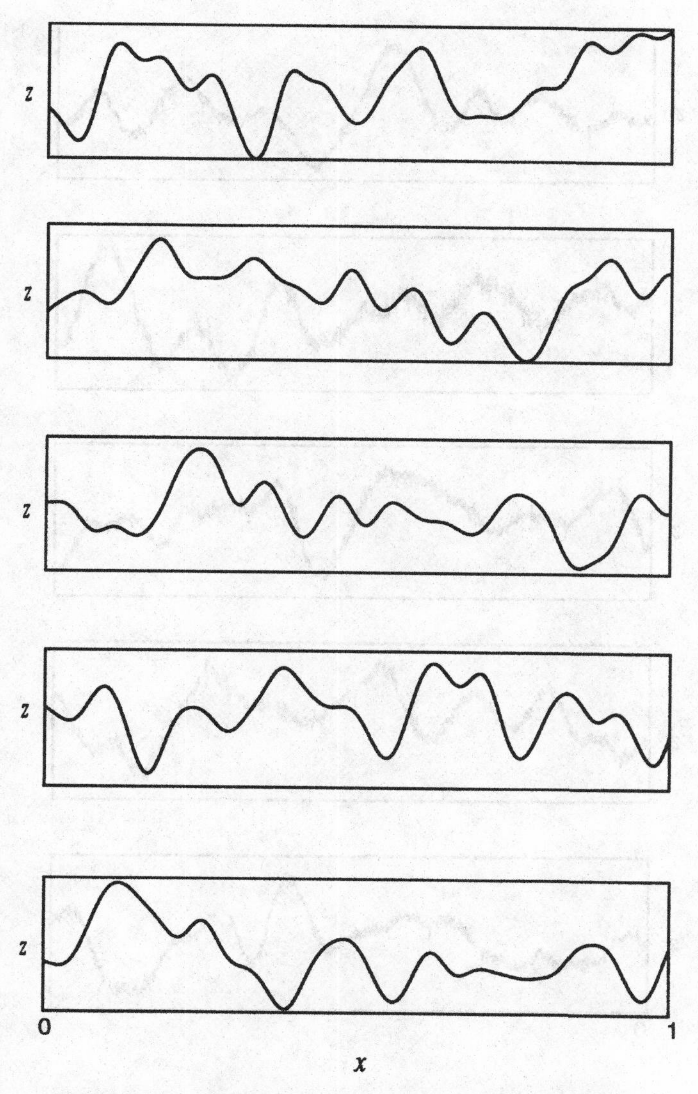

Figure 3.2 Five realizations from a family of $z(x)$ functions.

continuous but rough curves with ill-defined slopes. The curves in Figures 3.2 and 3.3 have fluctuations with much smaller periods than the fluctuations in Figure 3.4.

Assume for argument's sake that we have selected an ensemble and that we have computed the probability that a realization is the actual unknown, *i.e.*, we can specify that the probability that $z(\mathbf{x}) = z(\mathbf{x}; i)$, for any i, is P_i.

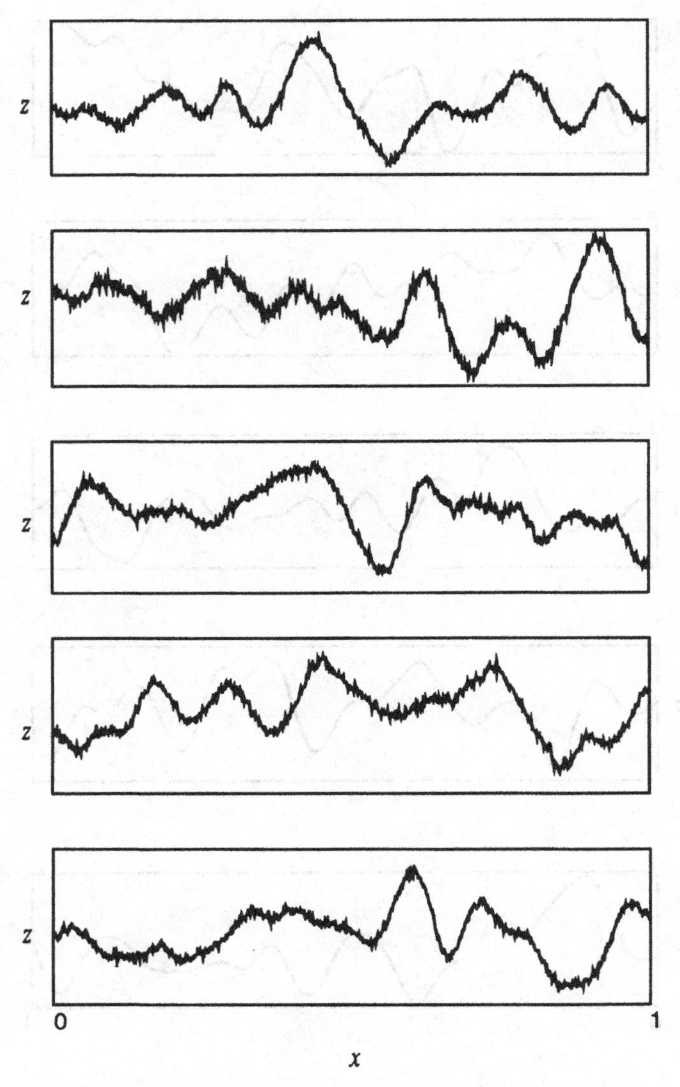

Figure 3.3 Five realizations from another family of functions.

Mathematically, we write

$$P_i = \Pr[z(\mathbf{x}) = z(\mathbf{x}; i)]. \tag{3.2}$$

(We will see later how we can assign these probabilities.)

The ensemble of realizations with their assigned probabilities defines what is known as a *random function* (or *random field* or *spatial stochastic process*). We

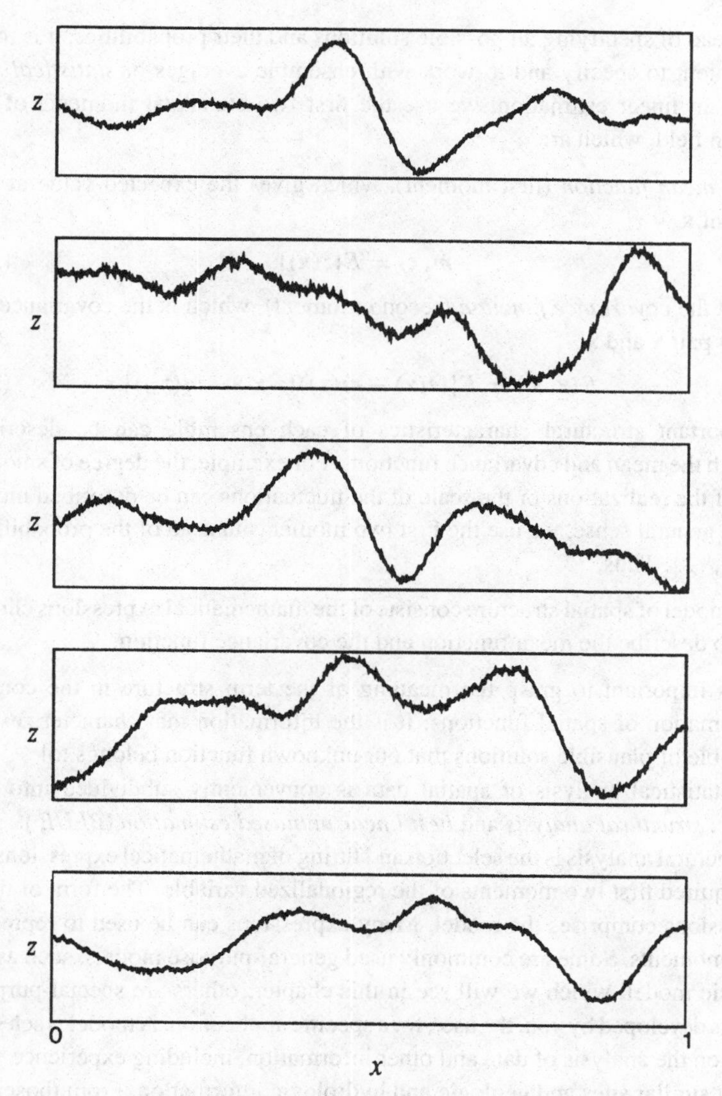

Figure 3.4 Five realizations from yet another family of functions.

are interested in calculating averages over all possible realizations. Expectation, denoted by the symbol E, is the process of computing a probability-weighted average over the ensemble. Thus, the expected value of z at location $\mathbf{x}$ is

$$E[z(\mathbf{x})] = P_1\, z(\mathbf{x}; 1) + P_2\, z(\mathbf{x}; 2) + \cdots = \sum_i P_i\, z(\mathbf{x}; i). \qquad (3.3)$$

Instead of specifying all possible solutions and their probabilities, it is more convenient to specify and to work with ensemble averages or *statistical moments*. In linear estimation, we use the first two statistical moments of the random field, which are

1. the *mean function* (first moment), which gives the expected value at any point $\mathbf{x}$,
$$m(\mathbf{x}) = E[z(\mathbf{x})], \tag{3.4}$$
2. and the *covariance function* (second moment), which is the covariance for any pair $\mathbf{x}$ and $\mathbf{x}'$,
$$R(\mathbf{x}, \mathbf{x}') = E[(z(\mathbf{x}) - m(\mathbf{x}))(z(\mathbf{x}') - m(\mathbf{x}'))]. \tag{3.5}$$

Important structural characteristics of each ensemble can be described through the mean and covariance functions. For example, the degree of smoothness of the realizations or the scale of the fluctuations can be described nicely.

In a general sense, we use the first two moments instead of the probabilities $P_1, P_2, \ldots$. Thus,

> the model of spatial structure consists of the mathematical expressions chosen to describe the mean function and the covariance function.

It is important to grasp the meaning of the term structure in the context of estimation of spatial functions: It is the information that characterizes the ensemble of plausible solutions that our unknown function belongs to!

A statistical analysis of spatial data is conveniently subdivided into two phases: *structural analysis* and *best linear unbiased estimation* (*BLUE*).

Structural analysis is the selection and fitting of mathematical expressions for the required first two moments of the regionalized variable. The form of these expressions comprises the model. Many expressions can be used to represent these moments. Some are commonly used general-purpose models, such as the intrinsic model, which we will see in this chapter; others are special-purpose models developed by you, the user, for a specific application. A model is selected based on the analysis of data and other information, including experience with data at similar sites and geologic and hydrologic information. From those, the analyst must decide whether the unknown function belongs, for example, in the ensemble of Figure 3.2, 3.3, or 3.4. Typically, model selection is an iterative procedure consisting of

(*a*) exploratory data analysis (see Chapter 2), on the basis of which a model is tentatively selected;

(*b*) parameter estimation, such as selection of numerical values for the parameters of the expressions of the mean and covariance function; and (*c*) model validation or diagnostic checking, which involves careful examination of the performance of the model in test cases.

Best linear unbiased estimation deals with taking into account specific observations. Specifically, we look for estimates that are as representative and accurate as possible, using the model developed during structural analysis and the specific observations.

The basic idea is that we proceed to figure out an unknown function (*e.g.*, the concentration over a cross section) in two stages.

1. During the first stage, structural analysis, the choice is narrowed down to the functions sharing certain characteristics, collectively known as structure.
2. During the second stage, the choice is narrowed down further by requiring that all possible solutions honor the data.

These ideas will become clearer after we study some examples. In this chapter, after Section 3.2, we will present one of the most commonly used geostatistical models, the intrinsic isotropic model.

3.2 Illustrative example

In this section, we will practice the concepts that we saw in the previous section by working on an example.

Consider the following family of functions:

$$z(x; u) = \sin(2\pi x + u), \tag{3.6}$$

where x is the spatial location (one-dimensional) and u is a random variable uniformly distributed between 0 and 2π. That is, the probability distribution of u is:

$$f(u) = \begin{cases} \dfrac{1}{2\pi}, & \text{if } 0 \leq u \leq 2\pi \\ 0, & \text{otherwise} \end{cases}. \tag{3.7}$$

We are asked to perform the following tasks:

(*a*) Justify why this information fully defines a random function $z(x)$.

(*b*) Compute the mean function $m(x)$.

(*c*) Compute the covariance function $R(x, x')$. Note that the covariance function depends only on $\|x - x'\|$, where $\| \ \|$ indicates the distance between x and x'. Plot the covariance function.

(*d*) Using a program that generates random variables (such as functions `rand` and `randn` in MATLAB) generate and plot five realizations of the random function.

(*e*) Now assume that you also measure the value 0.5 for z at $x = 0$. Conditional on this information: What are the possible solutions for $z(x)$ and their probabilities? What is the mean function? What is the covariance function?

The answers are given below:

(*a*) The random function $z(x)$ is fully defined because we have specified the way to generate all possible realizations or sample functions and the corresponding probability. For example, to generate a set of M equally likely realizations, generate M variates uniformly distributed between 0 and 2π (see Appendix C) and apply these in Equation (3.6).

(*b*) The mean at $z(x)$ is the weighted average over all values of z at this location. Intuitively, we expect that the mean will be zero. More systematically,

$$m(x) = E[\sin(2\pi x + u)] = \int_0^{2\pi} \sin(2\pi x + u)\frac{1}{2\pi}du = 0, \tag{3.8}$$

where $\frac{1}{2\pi}$ is the pdf (probability density function) of u in the interval between 0 and 2π. Thus, the mean is the same everywhere, 0. Note that the mean function is much simpler than any of the realizations.

(*c*) The covariance function is

$$\begin{aligned} R(x, x') &= E[\sin(2\pi x + u)\sin(2\pi x' + u)] \\ &= \int_0^{2\pi} \sin(2\pi x + u)\sin(2\pi x' + u)\frac{1}{2\pi}du \\ &= \frac{1}{2}\cos(2\pi(x - x')). \end{aligned} \tag{3.9}$$

Thus, we can see that the covariance function depends only on the distance $\|x - x'\|$. The variance, $R(x, x) = \frac{1}{2}$, is the same everywhere. See Figure 3.5.

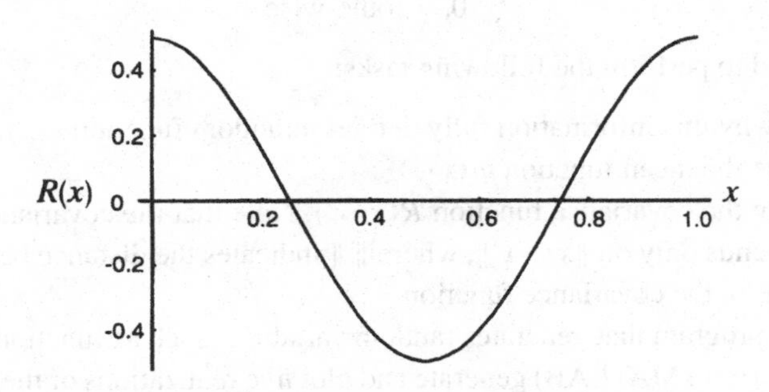

Figure 3.5 Plot of covariance function (periodic).

(*d*) We generate five random phases using the MATLAB command

```
2*pi*rand(5,1).
```

This picks five numbers between 0 and 2π. See Figure 3.6 for examples of sample functions.

(*e*) Given the available information, there are only two possible values for u: $\pi/6$ and $5\pi/6$; each is equally probable. Then, the only realizations that honor this measurement are

$$z_1(x) = \sin(2\pi x + \pi/6), \tag{3.10}$$

$$z_2(x) = \sin(2\pi x + 5\pi/6). \tag{3.11}$$

The conditional mean is

$$m_c(x) = \frac{1}{2}\sin(2\pi x + \pi/6) + \frac{1}{2}\sin(2\pi x + 5\pi/6) = \frac{1}{2}\cos(2\pi x). \tag{3.12}$$

See Figure 3.7. Note that the conditional mean function is more complex than the prior mean, which is zero everywhere. Also, note that the

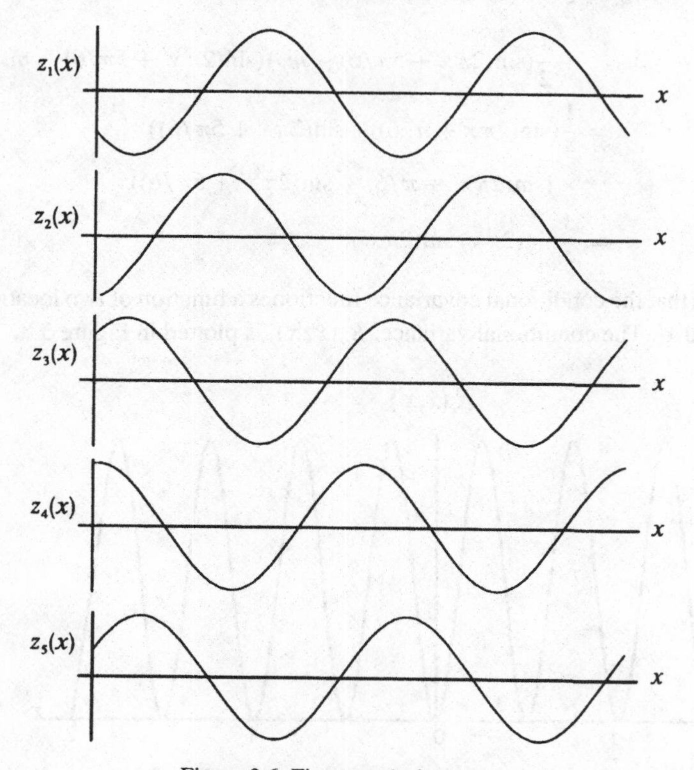

Figure 3.6 Five sample functions.

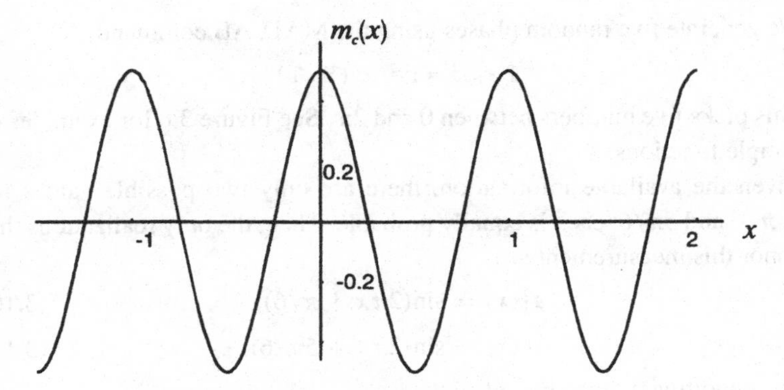

Figure 3.7 Conditional mean, given $z(0) = 0.5$.

conditional mean is smoother than either of the two possible solutions. The covariance function is

$$
\begin{aligned}
R_c(x, x') &= \frac{1}{2}(\sin(2\pi x + \pi/6) - m_c)(\sin(2\pi x' + \pi/6) - m_c) \\
&\quad + \frac{1}{2}(\sin(2\pi x + 5\pi/6) - m_c)(\sin(2\pi x' + 5\pi/6) - m_c) \\
&= \frac{1}{4}(\sin(2\pi x + \pi/6) - \sin(2\pi x + 5\pi/6)) \\
&\quad \times (\sin(2\pi x' + \pi/6) - \sin(2\pi x' + 5\pi/6)) \\
&= \frac{3}{4}\sin(2\pi x)\sin(2\pi x'). \qquad (3.13)
\end{aligned}
$$

Note that the conditional covariance function is a function of two locations, x and x'. The conditional variance, $R_c(x, x)$, is plotted in Figure 3.8.

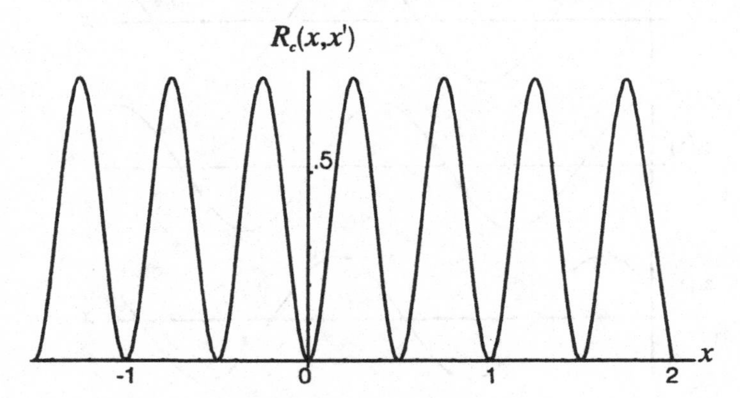

Figure 3.8 Conditional variance, given $z(0) = 0.5$.

Exercise 3.1 *Consider the random function z defined on one dimension x by*

$$z(x) = \cos(2\pi x + u_1) + \cos(\sqrt{2}\pi x + u_2),$$

where u_1 and u_2 are two independently distributed random variables with uniform distribution in the interval 0 to 2π.

(1) Find the mean function and the covariance function of z. (Plot results, if possible.)

(2) Generate and plot five realizations of this random function over the interval [0, 10].

3.3 Intrinsic isotropic model

In the outline of the general approach, it was mentioned that we need to write expressions for the mean and the covariance function. One of the simplest models is:

> The mean is constant and the two-point covariance function depends only on the distance between the two points.

That is,

$$E[z(\mathbf{x})] = m \tag{3.14}$$

and

$$E[(z(\mathbf{x}) - m)(z(\mathbf{x}') - m)] = R(h), \tag{3.15}$$

where

$$h = \|\mathbf{x} - \mathbf{x}'\| = \sqrt{(x_1 - x_1')^2 + (x_2 - x_2')^2 + (x_3 - x_3')^2} \tag{3.16}$$

is the distance between sampling locations $\mathbf{x}$ and $\mathbf{x}'$, $\mathbf{x} = [x_1, x_2, x_3]$, and $\mathbf{x}' = [x_1', x_2', x_3']$. Equations (3.14) and (3.15) comprise the *stationary* model; a random function $z(\mathbf{x})$ satisfying these conditions is called stationary.[1] This model is also *isotropic* because it uses only the length and not the orientation of the linear segment that connects the two points. In this chapter we will focus on isotropic models; we will see *anisotropic* models in other chapters.

The value of the covariance at $h = 0$ is known as the variance or the sill of the stationary function.

Exercise 3.2 *For the random function z(x) defined in Section 3.2:*

(1) Is z(x) stationary?

[1] Technically, this is known as "wide-sense" or "second-moment" stationary.

(2) Compute the ensemble average

$$\gamma(x, x') = \frac{1}{2}E[(z(x) - z(x'))^2].$$

(3) Show that $\gamma(x, x')$ depends only on the distance $\|x - x'\|$ and plot the expression $\gamma(\|x - x'\|)$ versus the separation distance $\|x - x'\|$. Show also that

$$\gamma(\|x - x'\|) + R(\|x - x'\|) = R(0)$$

is a constant value.

Mathematically speaking, there are several types of functions that satisfy Equations (3.14) and (3.15). The covariance function may be periodic (such as a cosine), aperiodic but consisting of a number of sinusoids, or a function that is none of the above but has a continuous power spectrum (from Fourier analysis) and finite variance. In estimation applications, it is the last type that is of interest so that, unless otherwise stated, we will assume that we deal with this type. All we need to know for now, without getting into Fourier analysis, is that for a stationary function (of the type we are interested in), the sill $R(0)$ is finite and the value $R(h)$ vanishes or tends to vanish when h exceeds a value called the *range*.

In order to apply this model in interpolation, we need to find the parameter m and to select an expression for the covariance function and find its parameters, such as the expression $R(h) = v \exp(-h/\ell)$ with parameters v and ℓ. Then, it is possible to *extrapolate* from the locations of the observations.

In most cases, the mean is not known beforehand but needs to be inferred from the data; to avoid this trouble, it may be more convenient to work with the *variogram*. The variogram is defined as

$$\gamma(h) = \frac{1}{2}E[(z(\mathbf{x}) - z(\mathbf{x}'))^2]. \tag{3.17}$$

(Originally, the term variogram was used for $2\gamma(h)$, and $\gamma(h)$ was called the *semivariogram*. Since we only use $\gamma(h)$, we will call it the variogram.) To underline the distinction between the experimental variogram, which is computed from the data, with the variogram, which is a mathematical expression, the latter is sometimes called the *theoretical variogram*.

For a stationary function, the relation between the variogram and the covariance function is

$$\begin{aligned}\gamma(h) &= \frac{1}{2}E[(z(\mathbf{x}) - z(\mathbf{x}'))^2] = \frac{1}{2}E[((z(\mathbf{x}) - m) - (z(\mathbf{x}') - m))^2] \\ &= -E[(z(\mathbf{x}) - m)(z(\mathbf{x}') - m)] + \frac{1}{2}E[(z(\mathbf{x}) - m)^2] \\ &\quad + \frac{1}{2}E[(z(\mathbf{x}') - m)^2] = -R(h) + R(0).\end{aligned} \tag{3.18}$$

That is, the variogram is minus the covariance function plus a constant (which happens to be the variance):

$$\gamma(h) = -R(h) + R(0). \tag{3.19}$$

Using the variogram, consider the model:

The mean is constant but unspecified and the two-point mean square difference depends only on the distance between the two locations.

That is,

$$E[z(\mathbf{x}) - z(\mathbf{x}')] = 0 \tag{3.20}$$

$$\frac{1}{2}E[(z(\mathbf{x}) - z(\mathbf{x}'))^2] = \gamma(h), \tag{3.21}$$

where $h = \|\mathbf{x} - \mathbf{x}'\|$ is the distance of the separation vector. Equations (3.20) and (3.21) comprise the *intrinsic isotropic model*.

At first, the reader may be unable to distinguish the intrinsic model from the stationary one. The difference is slight but important. Note, to begin with, that the stationary and intrinsic models differ in the parameters needed to characterize them as well as in mathematical generality. It takes less information to characterize the intrinsic model than the stationary model. Whereas both assume constant mean, in the intrinsic model we avoid ever using a numerical value for the mean. Furthermore, the stationary model may use the covariance function, which cannot be reconstructed only from the variogram over a distance smaller than the range. In a sense, to specify the covariance function one needs the variogram plus an extra number, the variance.

The intrinsic model is mathematically more inclusive (*i.e.*, general) than the stationary one. If $z(\mathbf{x})$ is stationary, then it is also intrinsic because Equation (3.20) follows from (3.14) and Equation (3.21) follows from (3.15) by using (3.19). For a stationary function, the variogram at large distances equals the sill, $\gamma(\infty) = R(0) = \sigma^2$. However, the important point is that not all intrinsic functions are stationary. As a practical rule, an intrinsic function is nonstationary if its variogram tends to infinity as h tends to infinity. For example, the

$$\gamma(h) = h \tag{3.22}$$

variogram characterizes an intrinsic function that is not stationary.

It is the intrinsic model that we will use in the remainder of this chapter. Invariably, the question asked is: What is the practical significance and meaning of this model? The practical significance is that it is a simple model that is useful in:

1. summarizing incomplete information and patterns in noisy data; and
2. allowing us to interpolate from observations of $z(\mathbf{x})$, as we will soon see.

The meaning of this model and its applicability range will be appreciated after we see examples and apply it to describe data.

Exercise 3.3 *If $z(x)$ is a stationary random function with mean m and covariance function $R(h)$, then find the mean and variance of*

$$\frac{1}{2}[z(\mathbf{x}) + z(\mathbf{x}')]$$

and

$$\frac{1}{2}[z(\mathbf{x}) - z(\mathbf{x}')],$$

where $\mathbf{x}$ and $\mathbf{x}'$ are two locations. Check each of these cases to see whether you can find the variance in terms of the variogram $\gamma(h) = R(0) - R(h)$ at separation distance $\|\mathbf{x} - \mathbf{x}'\|$, without necessarily specifying the variance of the process.

3.4 Common models

There are mathematical restrictions on which functions may be used as covariance functions or variograms. The reason is rather mundane: The variance of a linear combination of values of $z(\mathbf{x})$ at a number of points can be expressed in terms of covariance functions or variograms (see Exercise (3.3)); the models we use should be such that this expression cannot become negative. Criteria are discussed in references [21 and 30].

In practice, variograms describing the spatial structure of a function are formed by combining a small number of simple mathematically acceptable expressions or models. This section contains a list of such covariance functions R and variograms γ. Plots of some of them and sample functions are also presented. Symbol h stands for the distance between two points.

3.4.1 Stationary models

3.4.1.1 Gaussian model

For the Gaussian model we have

$$\begin{aligned} R(h) &= \sigma^2 \exp\left(-\frac{h^2}{L^2}\right) \\ \gamma(h) &= \sigma^2 \left(1 - \exp\left(-\frac{h^2}{L^2}\right)\right), \end{aligned} \tag{3.23}$$

where $\sigma^2 > 0$ and $L > 0$ are the two parameters of this model. Because the covariance function decays asymptotically, the range α is defined in practice as the distance at which the correlation is 0.05; *i.e.*, $\alpha \approx 7L/4$.

The Gaussian model is the only covariance in this list with parabolic behavior at the origin ($\gamma(h) \propto h^2$ for small h, where $\propto$ stands for "proportional to"), indicating that it represents a regionalized variable that is smooth enough to be differentiable (*i.e.*, the slope between two points tends to a well-defined limit as the distance between these points vanishes).

Figure 3.9 shows a sample function and a plot of the variogram γ and covariance function R for $\sigma^2 = 1$ and $L = 0.05$.

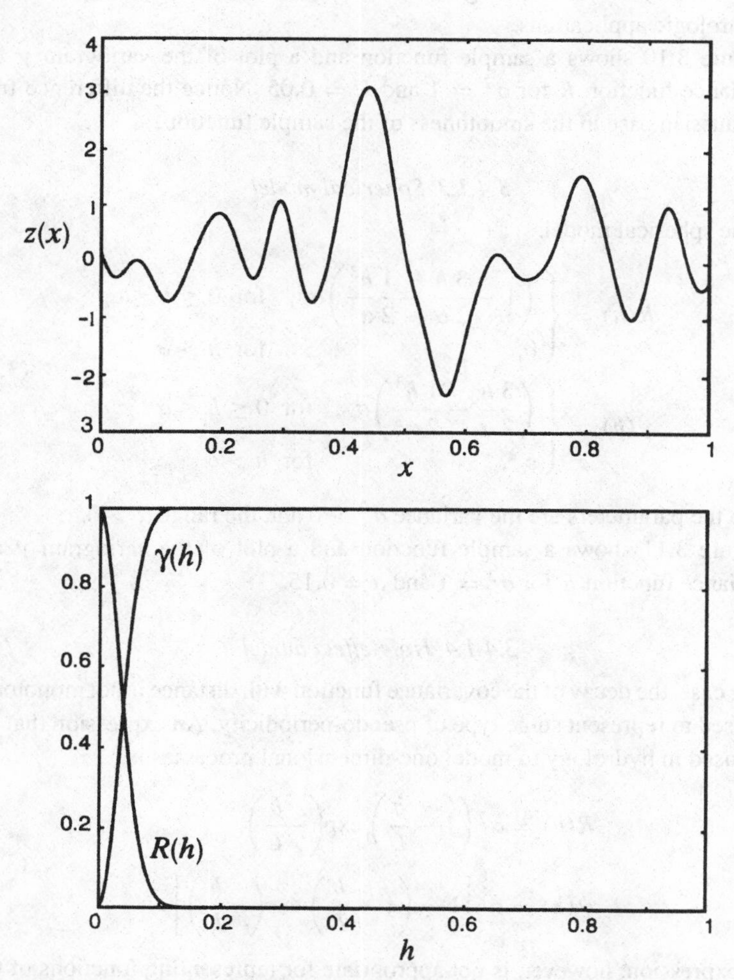

Figure 3.9 Sample function and model for Gaussian variogram and covariance function.

3.4.1.2 Exponential model

For this model, the covariance and variogram are given by

$$R(h) = \sigma^2 \exp\left(-\frac{h}{\ell}\right)$$
$$\gamma(h) = \sigma^2\left(1 - \exp\left(-\frac{h}{\ell}\right)\right), \tag{3.24}$$

where the parameters are the variance $\sigma^2 > 0$ and the length parameter (or *integral scale*) $\ell > 0$. The range is $\alpha \approx 3\ell$. This model is popular particularly in hydrologic applications.

Figure 3.10 shows a sample function and a plot of the variogram γ and covariance function R for $\sigma^2 = 1$ and $L = 0.05$. Notice the difference from the Gaussian case in the smoothness of the sample function.

3.4.1.3 Spherical model

For the spherical model,

$$R(h) = \begin{cases} \left(1 - \frac{3}{2}\frac{h}{\alpha} + \frac{1}{2}\frac{h^3}{\alpha^3}\right)\sigma^2, & \text{for } 0 \le h \le \alpha \\ 0, & \text{for } h > \alpha \end{cases}$$
$$\gamma(h) = \begin{cases} \left(\frac{3}{2}\frac{h}{\alpha} - \frac{1}{2}\frac{h^3}{\alpha^3}\right)\sigma^2, & \text{for } 0 \le h \le \alpha \\ \sigma^2, & \text{for } h > \alpha \end{cases}, \tag{3.25}$$

where the parameters are the variance $\sigma^2 > 0$ and the range $\alpha > 0$.

Figure 3.11 shows a sample function and a plot of the variogram γ and covariance function R for $\sigma^2 = 1$ and $a = 0.15$.

3.4.1.4 Hole-effect model

In this case, the decay of the covariance function with distance is not monotonic. It is used to represent some type of pseudo-periodicity. An expression that has been used in hydrology to model one-dimensional processes is

$$R(h) = \sigma^2\left(1 - \frac{h}{L}\right)\exp\left(-\frac{h}{L}\right)$$
$$\gamma(h) = \sigma^2\left[1 - \left(1 - \frac{h}{L}\right)\exp\left(-\frac{h}{L}\right)\right]. \tag{3.26}$$

This expression, however, is not appropriate for representing functions of two or three variables (dimensions).

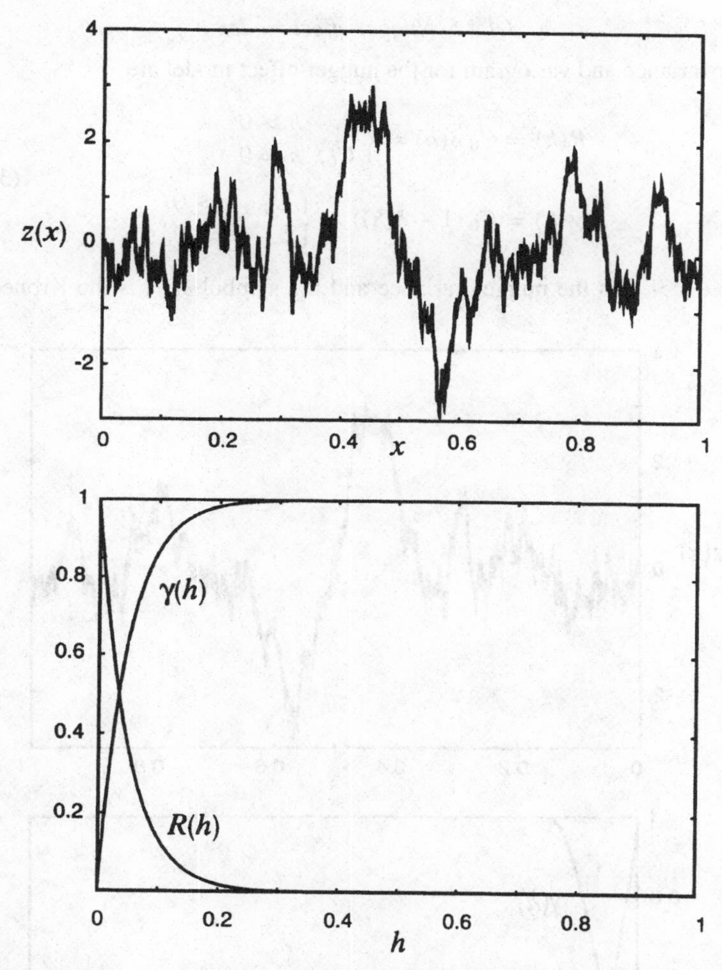

Figure 3.10 Sample function and model for exponential variogram and covariance function.

The hole-effect model describes processes for which excursions above the mean tend to be compensated by excursions below the mean. Figure 3.12 shows a sample function and a plot of the variogram γ and covariance function R for $\sigma^2 = 1$ and $L = 0.05$.

The exponential, spherical, and hole-effect models exhibit linear behavior at the origin, *i.e.*, $\gamma(h) \propto h$ for small h. The realizations of a random field with such a variogram are continuous but not differentiable, *i.e.*, they are less smooth than the realizations of a random field with a Gaussian covariance function.

3.4.1.5 Nugget-effect model

The covariance and variogram for the nugget-effect model are

$$
\begin{aligned}
R(h) &= C_0\,\delta(h) = \begin{cases} 0, & h > 0 \\ C_0, & h = 0 \end{cases} \\
\gamma(h) &= C_0\,(1 - \delta(h)) = \begin{cases} C_0, & h > 0 \\ 0, & h = 0 \end{cases},
\end{aligned}
\tag{3.27}
$$

where $C_0 > 0$ is the nugget variance and the symbol $\delta(h)$ is the Kronecker

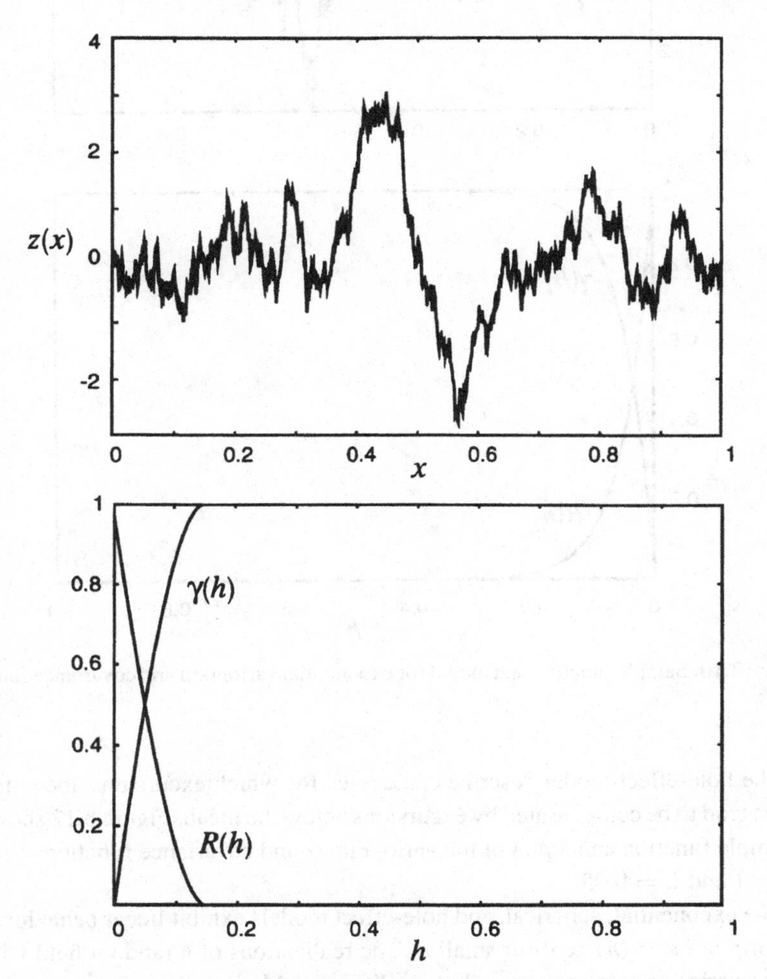

Figure 3.11 Sample function and model for spherical variogram and covariance function.

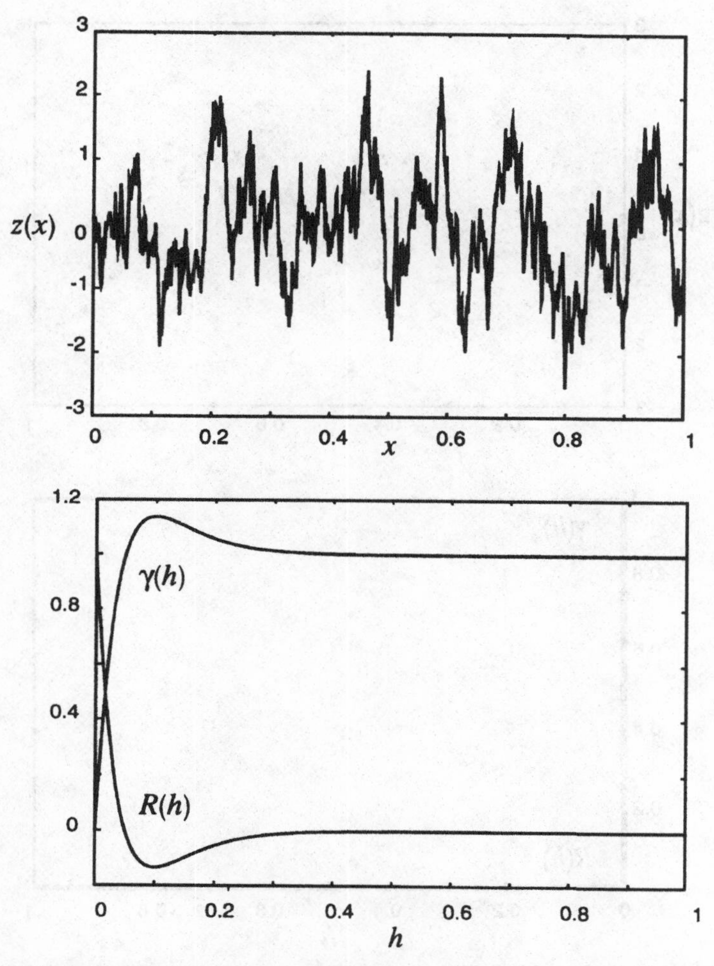

Figure 3.12 Sample function and model for hole-effect variogram and covariance function.

delta, which stands for 1 if $h = 0$ and for 0 in all other cases. The realizations of this random field are not continuous; *i.e.*, $z(\mathbf{x})$ can be different from $z(\mathbf{x}')$ no matter how small the distance $h = \|\mathbf{x} - \mathbf{x}'\|$ that separates them. Figure 3.13 shows a sample function and a plot of the variogram γ and covariance function R for $\sigma^2 = 1$ sampled at distances of about 0.004. Note that the sample function is discontinuous everywhere. The variogram and the covariance function are discontinuous at the origin. The variogram jumps up from 0 (at $h = 0$) to σ^2 (at $h > 0$); the covariance function drops off from σ^2 (at $h = 0$) to 0 (at $h > 0$).

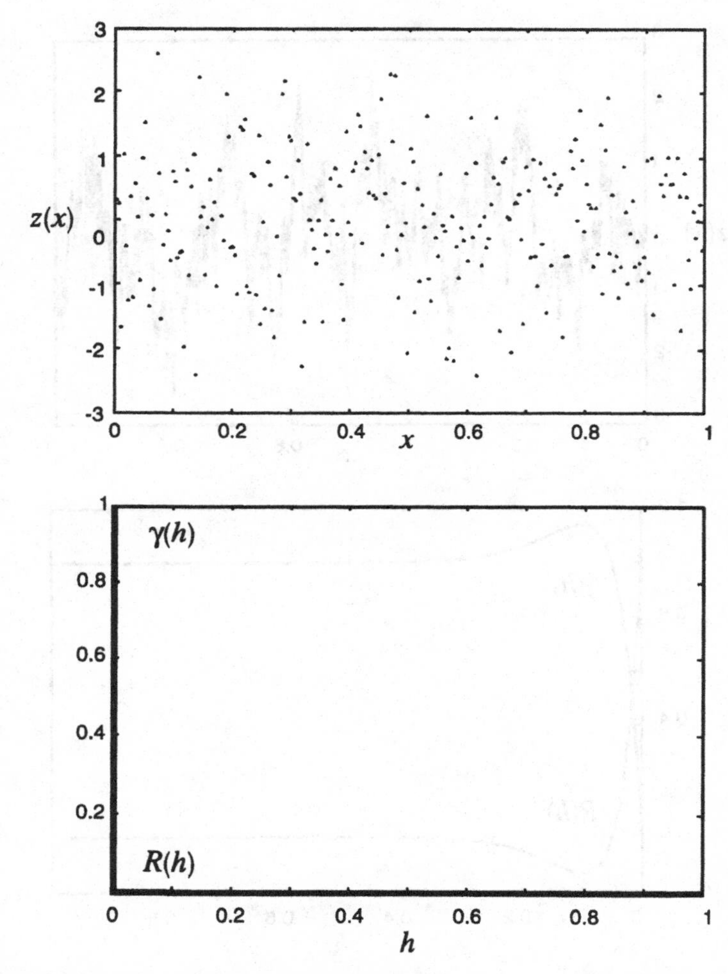

Figure 3.13 Sample function and model for nugget-effect variogram and covariance function.

The nugget-effect model represents microvariability in addition to random measurement error. Microvariability is variability at a scale smaller than the separation distance between the closest measurement points. For example, if rain gauges are located with a typical spacing of 1 km, then rainfall variability at the scale of 10 m or 100 m causes disparity among rainfall depths measured at the various gauges. You may think of the nugget semivariogram as a special case of the exponential semivariogram, $\gamma(h) = C_0[1 - \exp(-h/\ell)]$, when the typical distance h between observations is much larger than the inherent scale of the fluctuations of the phenomenon ℓ.

As already mentioned, discontinuous behavior can also be attributed to random measurement error, which produces observations that vary from gauge to gauge in an "unstructured" or "random" way.

Incidentally, the term "nugget" comes from mining, where the concentration of a mineral or the ore grade varies in a practically discontinuous fashion due to the presence of nuggets at the sampling points. Thus, nugget effect denotes variability at a scale shorter than the sampling interval.

3.4.2 Intrinsic nonstationary models

3.4.2.1 Power model

The variogram for the power model is

$$\gamma(h) = \theta \cdot h^s, \tag{3.28}$$

where the two parameters are the coefficient $\theta > 0$ and the exponent $0 < s < 2$. Figure 3.14 shows a sample function and a plot of the variogram γ for $\theta = 1$ and $s = 0.4$.

The power model describes a self-similar process: The realizations of such a process appear the same at any scale.

3.4.2.2 Linear model

In the linear model,

$$\gamma(h) = \theta h, \tag{3.29}$$

where the only parameter is the slope $\theta > 0$ of the variogram. Although it is a special case of the power model (for $s = 1$), we mention it separately because of its usefulness in applications. Figure 3.15 shows a sample function and a plot of the variogram γ for $\theta = 1$.

Note that in all of the above models, $\gamma(0) = 0$.

3.4.2.3 Logarithmic model

The variogram for the logarithmic model is

$$\gamma(h) = A \log(h), \tag{3.30}$$

where $A > 0$. This model can be used only for integrals over finite volumes and cannot be used directly with point values of the regionalized variable. For example, it can be used to estimate solute mass over a finite volume given measurements of mass in samples also of finite volume.

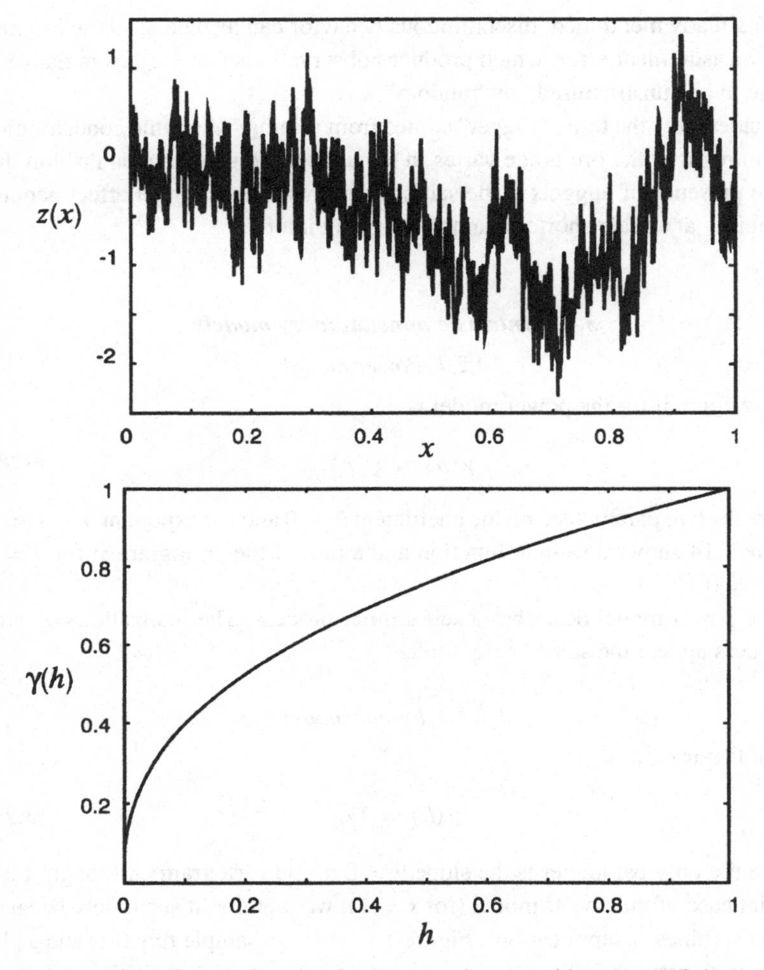

Figure 3.14 Sample function for power variogram with exponent 0.4.

3.4.3 Model superposition

By adding variograms of types listed in this section, one can obtain other mathematically acceptable variograms. For example, combining the linear and nugget-effect semivariograms we obtain another useful model,

$$\gamma(h) = \begin{cases} C_0 + \theta h, & h > 0 \\ 0, & h = 0 \end{cases} \tag{3.31}$$

with two parameters, $C_0 \geq 0$ and $\theta \geq 0$. In this fashion, one can find a model

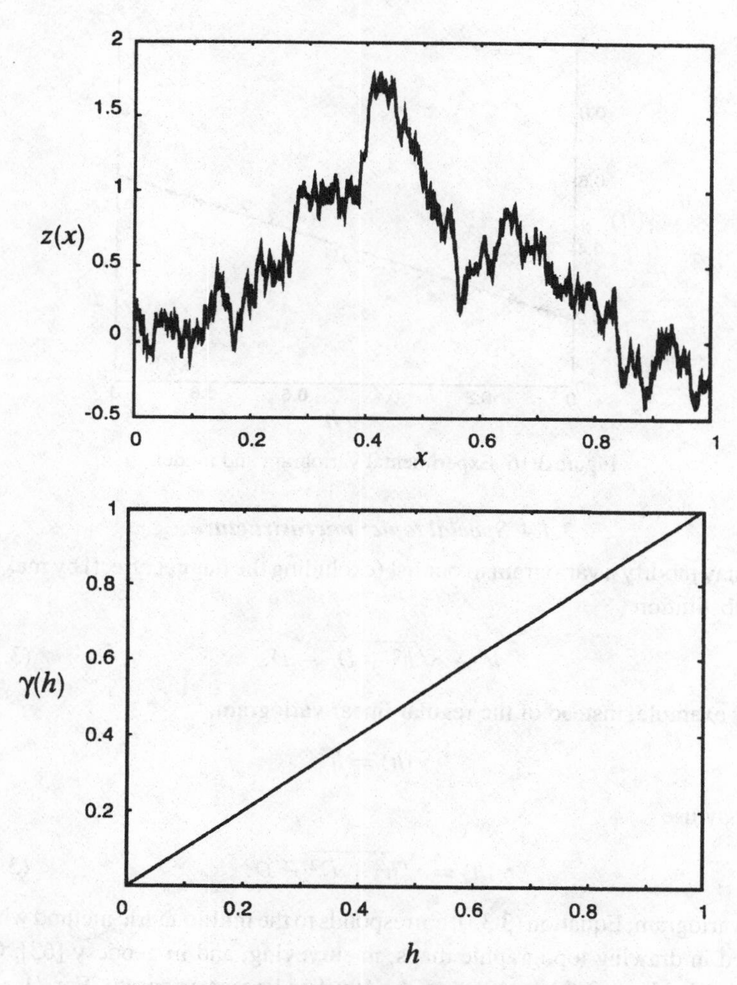

Figure 3.15 Sample function for linear variogram.

that adequately represents the structure of a spatial variable. For example, one may fit Equation (3.31) to the experimental variogram, as shown in Figure 3.16.

Exercise 3.4 *Consider the following three variogram models:*

1. $R(h) = \exp(-h^2/0.01) + 0.002\exp(-h/0.0025)$
2. $R(h) = \exp(-h^2/0.0025) + 0.01\exp(-h/0.0025)$
3. $R(h) = \exp(-h^2/0.0025)$.

Each of these models corresponds to one ensemble from those sampled in Figures 3.2, 3.3, and 3.4. Find which model corresponds to each figure.

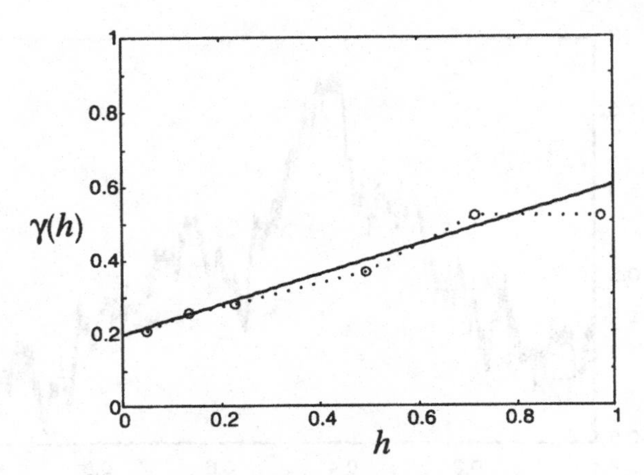

Figure 3.16 Experimental variogram and model.

3.4.4 Special topic: microstructure

One may modify a variogram in our list (excluding the nugget effect) by making the substitution

$$h \rightarrow \sqrt{h^2 + D^2} - D. \tag{3.32}$$

For example, instead of the regular linear variogram,

$$\gamma(h) = h$$

one may use

$$\gamma(h) = \sqrt{h^2 + D^2} - D. \tag{3.33}$$

This variogram, Equation (3.33), corresponds to the multiquadric method which is used in drawing topographic maps, in surveying, and in geodesy [63]. Obviously, for $D = 0$ this reduces to the familiar linear variogram. For $D > 0$, however, the interpolation at the observation points becomes smooth (differentiable) at the observation points. Intuitively, D is a measure of the radius of curvature of the curve or surface at the observation points, or "microstructure." This approach is valued by some practitioners because it produces smooth and aesthetically pleasing maps.

The substitution of Equation (3.32) allows us to introduce models with two different scale parameters. For example, the modified exponential variogram,

$$\gamma(h) = \sigma^2 \left(1 - \exp\left(-\frac{\sqrt{h^2 + D^2} - D}{l} \right) \right), \tag{3.34}$$

has two scale parameters: l, which controls the large-scale correlation in the function, and D, which controls the curvature of the function (assuming that $l \gg D$).

3.5 Interpolation using kriging

Consider, for example, some log-transmissivity measurements and assume that the intrinsic isotropic model is tentatively selected. Next, we need to specify the variogram. One may use the experimental variogram to fit the expression for the model variogram. To enure that the model is mathematically acceptable, the exponential variogram may be chosen from one of the models listed in Section 3.4, and its parameters may be adjusted to approximate the experimental variogram. At the end, the following mathematical expression is used:

$$\gamma(h) = v\left[1 - \exp\left(-\frac{h}{\ell}\right)\right], \tag{3.35}$$

where h is the separation distance and v and ℓ are coefficients. It will be seen later how we can complete the structural analysis by returning to address the questions of whether this is a good model and, if so, whether we can refine it by choosing better values for the parameters. (The details of variogram analysis are dealt with in the next chapter.) For now, accept this model and proceed to solve an interpolation problem using kriging.

Before getting to the kriging equations, let us discuss what we try to accomplish. Selection of a variogram is equivalent to selecting an ensemble of solutions; we assume that the function we seek can be found in this collection of functions. The variogram and the ensemble it defines were found on the basis of considerations of spatial structure. However, the ensemble contains functions that do not "honor" the measurements. For example, if we have measured and (disregarding measurement error) thus know the value of z at a certain location, it is only reasonable to eliminate from further consideration the functions that do not have the right value at this location.

That is the basic idea. In practice, we do not keep track of individual realizations but we try to compute a mean value and a variance of the ensemble of the functions that are left after we account for the information about the observations. These are the conditional mean (or *best estimate*) and variance (or *mean square error of estimation*) of the unknown function $z(\mathbf{x})$. To facilitate the analysis, we limit our attention to best estimates that depend on the data *linearly*.

Kriging involves applying the general methodology known as best linear unbiased estimation (BLUE) to intrinsic functions. The theory of kriging as well as its implementation on a computer are straightforward, despite the intimidating equations that it involves. We will consider the generic problem:

Given n measurements of z at locations with spatial coordinates $\mathbf{x}_1, \mathbf{x}_2, \ldots, \mathbf{x}_n$, estimate the value of z at point $\mathbf{x}_0$.

An estimator is simply a procedure or formula that uses data to find a representative value, or estimate, of the unknown quantity. We limit our attention to estimators that are linear combinations of the measurements, *i.e.*,

$$\hat{z}_0 = \sum_{i=1}^{n} \lambda_i z(\mathbf{x}_i). \tag{3.36}$$

Thus, the problem is reduced to selecting a set of coefficients $\lambda_1, \ldots, \lambda_n$.

The difference between the estimate $\hat{z}_0$ and the actual value $z(\mathbf{x}_0)$ is the *estimation error*:

$$\hat{z}_0 - z(\mathbf{x}_0) = \sum_{i=1}^{n} \lambda_i z(\mathbf{x}_i) - z(\mathbf{x}_0). \tag{3.37}$$

Obviously, we seek a good estimator. Specifically, we select the coefficients so that the estimator meets the following specifications:

Unbiasedness On the average (*i.e.*, over all possible solutions or realizations), the estimation error must be zero. That is,

$$E[\hat{z}_0 - z(\mathbf{x}_0)] = \sum_{i=1}^{n} \lambda_i m - m = \left(\sum_{i=1}^{n} \lambda_i - 1 \right) m = 0. \tag{3.38}$$

But the numerical value of the mean, m, is not specified. For the estimator to be unbiased for *any* value of the mean, it is required that

$$\sum_{i=1}^{n} \lambda_i = 1. \tag{3.39}$$

Imposing the unbiasedness constraint eliminates the unknown parameter m.

Minimum Variance The mean square estimation error must be minimum. We can compute the mean square error in terms of the variogram if we use condition (3.39). After some algebra, we get

$$E[(\hat{z}_0 - z(\mathbf{x}_0))^2] = -\sum_{i=1}^{n}\sum_{j=1}^{n} \lambda_i \lambda_j \gamma(\|\mathbf{x}_i - \mathbf{x}_j\|) + 2\sum_{i=1}^{n} \lambda_i \gamma(\|\mathbf{x}_i - \mathbf{x}_0\|). \tag{3.40}$$

Thus the problem of best (minimum mean-square error) unbiased estimation of the λ coefficients may be reduced to the constrained optimization problem:

Select the values of $\lambda_1, \ldots, \lambda_n$ that minimize the expression (3.40) while satisfying constraint (3.39). Expression (3.40) is called the objective function, because we try to minimize it; Equation (3.39) is called a constraint, because it poses a restriction on the values that we may assign to the coefficients.

3.6 Kriging system

In the previous section we formulated a constrained optimization problem with a quadratic objective function and a linear constraint. This problem can be solved easily using Lagrange multipliers, a standard optimization method (see Appendix B); the necessary conditions for the minimization are given by the linear *kriging system* of $n+1$ equations with $n+1$ unknowns:

$$-\sum_{j=1}^{n} \lambda_j \gamma(\|\mathbf{x}_i - \mathbf{x}_j\|) + \nu = -\gamma(\|\mathbf{x}_i - \mathbf{x}_0\|), \quad i = 1, 2, \ldots, n \tag{3.41}$$

$$\sum_{j=1}^{n} \lambda_j = 1, \tag{3.42}$$

where ν is a Lagrange multiplier.

It is common practice to write the kriging system in matrix notation. Let $\mathbf{x}$ be the vector of the unknowns

$$\mathbf{x} = \begin{bmatrix} \lambda_1 \\ \lambda_2 \\ \vdots \\ \lambda_n \\ \nu \end{bmatrix}, \tag{3.43}$$

$\mathbf{b}$ the right-hand-side vector

$$\mathbf{b} = \begin{bmatrix} -\gamma\left(\|\mathbf{x}_1 - \mathbf{x}_0\|\right) \\ -\gamma\left(\|\mathbf{x}_2 - \mathbf{x}_0\|\right) \\ \vdots \\ -\gamma\left(\|\mathbf{x}_n - \mathbf{x}_0\|\right) \\ 1 \end{bmatrix}, \tag{3.44}$$

and $\mathbf{A}$ the matrix of coefficients

$$\mathbf{A} = \begin{bmatrix} 0 & -\gamma(\|\mathbf{x}_1 - \mathbf{x}_2\|) & \cdots & -\gamma(\|\mathbf{x}_1 - \mathbf{x}_n\|) & 1 \\ -\gamma(\|\mathbf{x}_2 - \mathbf{x}_1\|) & 0 & \cdots & -\gamma(\|\mathbf{x}_2 - \mathbf{x}_n\|) & 1 \\ \vdots & \vdots & & \vdots & \vdots \\ -\gamma(\|\mathbf{x}_n - \mathbf{x}_1\|) & -\gamma(\|\mathbf{x}_n - \mathbf{x}_2\|) & \cdots & 0 & 1 \\ 1 & 1 & \cdots & 1 & 0 \end{bmatrix}. \tag{3.45}$$

We denote by A_{ij} the element of $\mathbf{A}$ at the i-th row and j-th column, and by x_i and b_i the element at the i-th row of $\mathbf{x}$ and $\mathbf{b}$, respectively. Notice that $\mathbf{A}$ is symmetric, *i.e.*, $A_{ij} = A_{ji}$.

The kriging system can be written as

$$\sum_{j=1}^{n+1} A_{ij}x_j = b_i, \quad \text{for } i = 1, 2, \ldots, n+1 \tag{3.46}$$

or, in matrix notation,

$$\mathbf{Ax} = \mathbf{b}.$$

Note that the value of γ vanishes at exactly zero separation distance. Thus, on the left-hand side, the diagonal elements of the matrix of the coefficients should be taken equal to zero.

All BLUE problems boil down to the solution of a linear system such as this. Computer programs for their solution are readily available. Solving this system, we obtain $\lambda_1, \lambda_2, \ldots, \lambda_n, \nu$. In this manner, the linear estimator of Equation (3.36) is fully specified. Furthermore, we can quantify the accuracy of the estimate through the mean square estimation error. The mean square estimation error may be obtained by substituting in Equation (3.40) the values of $\lambda_1, \lambda_2, \ldots, \lambda_n$ obtained from the solution of the kriging system.

Actually, using the already computed Lagrange multiplier ν we can find a computationally more efficient method. For the optimal values of the kriging coefficients, substituting

$$-\sum_{j=1}^{n} \lambda_j \gamma(\|\mathbf{x}_i - \mathbf{x}_j\|) = -\nu - \gamma(\|\mathbf{x}_i - \mathbf{x}_0\|) \tag{3.47}$$

in (3.40) we obtain the useful expression

$$\sigma_0^2 = E[(\hat{z}_0 - z(\mathbf{x}_0))^2] = -\nu + \sum_{i=1}^{n} \lambda_i \gamma\left(\|\mathbf{x}_i - \mathbf{x}_0\|\right). \tag{3.48}$$

In many cases we are interested in obtaining the 95% confidence interval of estimation of $z(x_0)$. This is the interval that contains the actual value of $z(x_0)$ with probability 0.95. Calculation of this interval is not possible without making some explicit assumptions about the probability distribution of $z(x)$. Assuming normality, *i.e.*, Gaussian distribution of the estimation errors, we get a confidence interval approximately $[\hat{z}_0 - 2\sigma_0, \hat{z}_0 + 2\sigma_0]$.[2]

[2] Even when normality is not explicitly mentioned, this is the conventional way of obtaining confidence intervals in linear geostatistics.

Example 3.1 *Consider a one-dimensional function with variogram $\gamma(h) = 1 + h$, for $h > 0$, and three measurements, at locations $x_1 = 0$, $x_2 = 1$, and $x_3 = 3$. Assume that we want to estimate the value of the function in the neighborhood of the measurement points, at location $x_0 = 2$. The kriging system of equations, which gives the weights λ_1, λ_2, and λ_3, is*

$$\begin{array}{rrrrcr} & -2\lambda_2 & -4\lambda_3 & +\nu & = & -3 \\ -2\lambda_1 & & -3\lambda_3 & +\nu & = & -2 \\ -4\lambda_1 & -3\lambda_2 & & +\nu & = & -2 \\ \lambda_1 & +\lambda_2 & +\lambda_3 & & = & 1 \end{array}$$

or, in matrix notation,

$$\begin{bmatrix} 0 & -2 & -4 & 1 \\ -2 & 0 & -3 & 1 \\ -4 & -3 & 0 & 1 \\ 1 & 1 & 1 & 0 \end{bmatrix} \begin{bmatrix} \lambda_1 \\ \lambda_2 \\ \lambda_3 \\ \nu \end{bmatrix} = \begin{bmatrix} -3 \\ -2 \\ -2 \\ 1 \end{bmatrix}.$$

The mean square estimation error is

$$-\nu + 3\lambda_1 + 2\lambda_2 + 2\lambda_3.$$

Solving the system, we obtain $\lambda_1 = 0.1304$, $\lambda_2 = 0.3913$, $\lambda_3 = 0.4783$, and $\nu = -0.304$. The mean square estimation error is $MSE = 2.43$.

For $x_0 = 0$, i.e., a position coinciding with an observation point, the kriging system is

$$\begin{bmatrix} 0 & -2 & -4 & 1 \\ -2 & 0 & -3 & 1 \\ -4 & -3 & 0 & 1 \\ 1 & 1 & 1 & 0 \end{bmatrix} \begin{bmatrix} \lambda_1 \\ \lambda_2 \\ \lambda_3 \\ \nu \end{bmatrix} = \begin{bmatrix} 0 \\ -2 \\ -4 \\ 1 \end{bmatrix}.$$

By inspection, one can verify that the only possible solution is $\lambda_1 = 1$, $\lambda_2 = \lambda_3 = 0$, and $\nu = 0$. The mean square estimation error is $MSE = 0$, indicating error-free prediction. The kriging estimate at a measurement location is the measurement itself.

Example 3.2 *We will illustrate how we can program kriging using MATLAB with the following example.*

We have two observations: 1.22 at location (9.7, 47.6) and 2.822 at (43.8, 24.6). Find best estimate and MSE at (18.8, 67.9) if the variogram is

$$\gamma(h) = \begin{cases} 0.1 + 0.006h, & h > 0 \\ 0, & h = 0 \end{cases}.$$

The input data are

$$zdata = \begin{bmatrix} 1.22 \\ 2.822 \end{bmatrix}, \; xdata = \begin{bmatrix} 9.7 & 47.6 \\ 43.8 & 24.6 \end{bmatrix}, \; x0 = [18.8 \;\; 67.9],$$

from which MATLAB yeilds the following results:

$$A = \begin{bmatrix} 0 & -0.3468 & 1 \\ -0.3468 & 0 & 1 \\ 1 & 1 & 0 \end{bmatrix}, \; b = \begin{bmatrix} -0.2335 \\ -0.4000 \\ 1 \end{bmatrix}$$

$$coef = \begin{bmatrix} 0.7401 \\ 0.2599 \\ -0.1433 \end{bmatrix}, \; z0 = 1.6364, \; MSE = 0.4201.$$

The MATLAB program used in this example was:

```
n=2  %number of observations
zdata=[1.22; 2.822]  %column of observations
xdata=[9.7,47.6;43.8,24.6]  %column of obs.
      %coordinates
x0=[18.8,67.9]  %location of unknown
A=zeros(n+1,n+1); b=zeros(n+1,1);  %initialize
%A is matrix of coefficients and
%b is right-hand-side vector of kriging system
A(n+1,1)=1; A(1,n+1)=1;
for i=2:n
  A(n+1,i)=1; A(i,n+1)=1;
  for j=1:i-1
    A(i,j)=-.1-0.006*norm(xdata(i,:)-xdata(j,:));
    A(j,i)=A(i,j);
  end
end
for i=1:n
  b(i)=-.1-0.006*norm(xdata(i,:)-x0);
end
b(n+1)=1;
A,b
coef=A\b  %solve kriging system
z0=coef(1:n)'*zdata  %Find best estimate
MSE=-b'*coef  %find MSE
```

Exercise 3.5 *You are given n measurements of concentration $z(x_1), \ldots, z(x_n)$ in a borehole at distinct depths $x_1 < \cdots < x_n$. The concentration is modeled as an intrinsic function with linear variogram*

$$\gamma(\|x - x'\|) = a\|x - x'\|,$$

where a is a positive parameter. Write the kriging system for the estimation of z at location x_0. Then, show that the only coefficients that are not zero are the ones that correspond to the two adjacent measurement points, i.e., the ones between which x_0 is located. (Hint: Because the kriging system has a unique solution, all you need to do is verify that this solution satisfies the kriging system.) Write the expressions for the values of these coefficients and the mean square error. Draw sketches depicting how the best estimate and the mean square error vary between the adjacent measurement points.

3.7 Kriging with moving neighborhood

In some applications, instead of using all n observations at the same time, only a subset is selected in the neighborhood of the estimation point $\mathbf{x}_0$. For example, one may use only the 20 observations that are positioned closest to $\mathbf{x}_0$, or one may use all observations within a specified radius, or one may use a combination of these strategies. As the estimation point moves, the set of observations used in estimation change with it.

One motivation for using such a "moving neighborhood" has been the reduction of computational cost achieved by solving kriging systems with fewer unknowns. However, in many applications there are fewer than, say, 300 observations, and the cost of solving a system of that size in a workstation circa 1996 is insignificant (particularly in comparison to the cost of collecting the data). Consequently, except in the case of very large data sets, computational convenience should not be the overriding reason for choosing a moving neighborhood. In Section 3.10, we point out that in kriging for contouring it may be advantageous to use all observations at once.

Another motivation has been to make the estimate at $\mathbf{x}_0$ dependent only on observations in its neighborhood, which is often a desirable characteristic. However, the same objective can be achieved by using all data with an appropriate variogram (such as the linear one) that assigns very small weights to observations at distant points. If the weights corresponding to points near the border of the neighborhood are not small and the moving neighborhood method is applied in contouring, the estimated surface will have discontinuities that are unsightly as well as unreasonable, if they are the artifact of the arbitrary selection of a moving neighborhood.

In conclusion, using a moving neighborhood is not recommended as a general practice (although it may be useful in special cases) and will not be discussed further in this book.

3.8 Nonnegativity

Quite often z, the function of interest, is known to be nonnegative. For example, z may represent transmissivity or storage coefficient or solute concentration. However, kriging does not account for this nonnegativity requirement. An indication that nonnegativity may need to be enforced is when the 95% confidence interval appears unrealistic because it includes negative values.

The most practical way to enforce nonnegativity is to make a one-to-one variable transformation. The prevalent transformation is the logarithmic

$$y(\mathbf{x}) = \ln(z(\mathbf{x})). \tag{3.49}$$

Then, we proceed as follows.

1. We transform the data and develop a variogram for y.
2. We perform kriging to obtain best estimates and 95% confidence interval for $y(\mathbf{x}_0)$, $\hat{y}_0$, and $[y_l, y_u]$.
3. Through the backtransformation

$$z(\mathbf{x}) = \exp(y(\mathbf{x})) \tag{3.50}$$

   we obtain a best estimate $\exp(\hat{y}_0)$ and confidence interval $[\exp(y_l), \exp(y_u)]$ for $z(\mathbf{x}_0)$.

Note, however, that $\exp(\hat{y}_0)$ is a best estimate not in the sense of minimum mean square error but in the sense that it is the median. (This does not represent a difficulty because mean square error is not a particularly useful measure for asymmetric distributions.)

One may also use other transformations that are less "drastic" than the logarithmic. A useful model is the power transformation encountered in Chapter 2, Equation (2.10), which includes the logarithmic transformation as a special case. That is, we may apply Equation (2.10) instead of the logarithmic and then

$$z(\mathbf{x}) = (\kappa y(\mathbf{x}) + 1)^{\frac{1}{\kappa}} \tag{3.51}$$

is used for the backtransformation. However, care should be taken to select a value of parameter κ small enough to deal with values of y that satisfy the requirement

$$\kappa y(\mathbf{x}) + 1 \geq 0 \tag{3.52}$$

or else the transformation is not monotonic and the results may be nonsensical.

3.9 Derivation

This section contains a step-by-step derivation of the expression for the mean square error, Equation (3.40). This is an exercise in algebra, really, and you may decide to skip it; however, the method is worth learning because the same steps are followed to derive any best linear unbiased estimator.

It is convenient to work with the covariance function $R(h)$, instead of the variogram; so let us assume that the covariance function exists. (This is the case for stationary functions. However, the result applies even for nonstationary intrinsic functions, as we will see, because ultimately the result depends only on the variogram.) Then

$$E[(\hat{z}_0 - z(\mathbf{x}_0))^2] = E\left[\left(\sum_{i=1}^{n} \lambda_i z(\mathbf{x}_i) - z(\mathbf{x}_0)\right)^2\right].$$

First, we add and subtract the mean (making use of the unbiasedness constraint, Equation (3.39)) to get

$$E\left[\left(\sum_{i=1}^{n} \lambda_i (z(\mathbf{x}_i) - m) - (z(\mathbf{x}_0) - m)\right)^2\right].$$

Expanding the square yields

$$E\left[\left(\sum_{i=1}^{n} \lambda_i (z(\mathbf{x}_i) - m)\right)^2 + (z(\mathbf{x}_0) - m)^2 - 2\sum_{i=1}^{n} \lambda_i (z(\mathbf{x}_i) - m)(z(\mathbf{x}_0) - m)\right].$$

Replacing the square of simple summation with double summation, we have

$$E\sum_{i=1}^{n}\sum_{j=1}^{n} \lambda_i \lambda_j (z(\mathbf{x}_i) - m)(z(\mathbf{x}_j) - m) + (z(\mathbf{x}_0) - m)^2$$
$$- 2\sum_{i=1}^{n} \lambda_i (z(\mathbf{x}_i) - m)(z(\mathbf{x}_0) - m),$$

which, upon changing the order of expectation and summation, becomes

$$\sum_{i=1}^{n}\sum_{j=1}^{n} \lambda_i \lambda_j E[(z(\mathbf{x}_i) - m)(z(\mathbf{x}_j) - m)] + E[(z(\mathbf{x}_0) - m)^2]$$
$$- 2\sum_{i=1}^{n} \lambda_i E[(z(\mathbf{x}_i) - m)(z(\mathbf{x}_0) - m)].$$

This equation can be expressed using the covariance function as

$$\sum_{i=1}^{n}\sum_{j=1}^{n} \lambda_i \lambda_j R(\|\mathbf{x}_i - \mathbf{x}_j\|) + R(0) - 2\sum_{i=1}^{n} \lambda_i R(\|\mathbf{x}_i - \mathbf{x}_0\|).$$

Finally, we substitute using $R(h) = R(0) - \gamma(h)$ and use the fact that the λs sum to 1 to obtain

$$-\sum_{i=1}^{n}\sum_{j=1}^{n}\lambda_i\lambda_j\gamma(\|\mathbf{x}_i - \mathbf{x}_j\|) + 2\sum_{i=1}^{n}\lambda_i\gamma(\|\mathbf{x}_i - \mathbf{x}_0\|), \tag{3.53}$$

which is (3.40).

3.10 The function estimate

The "textbook" version of kriging is to formulate the best estimate of the function at a point as a linear weighting of the data, then to determine the weights by solving a system of linear equations of order equal to the number of observations plus one. However, in practice, one is often interested in estimating values on the nodes of a fine mesh for various reasons, *e.g.*, in order to display the best estimates using computer graphics. To apply the kriging technique (*i.e.*, to solve a system of equations, independently for every node) is inefficient in the case of fine grids with many nodes, because solving a system of n equations involves operations of the order of $\frac{n^3}{3}$.

We can, however, write analytical expressions for the best estimate at any point $\mathbf{x}$, as discussed in references [30 and 86]. The method is as follows: Set

$$\hat{z}\,(\mathbf{x}) = -\sum_{j=1}^{n}\gamma(\|\mathbf{x} - \mathbf{x}_j\|)\xi_j + \hat{\beta}, \tag{3.54}$$

where the ξ and $\hat{\beta}$ coefficients are found from the solution of a single system of $n + 1$ equations with $n + 1$ unknowns:

$$\begin{aligned} -\sum_{j=1}^{n}\gamma(\|\mathbf{x}_i - \mathbf{x}_j\|)\xi_j + \hat{\beta} &= z(\mathbf{x}_i), \quad \text{for } i = 1, \ldots, n \\ \sum_{j=1}^{n}\xi_j &= 0. \end{aligned} \tag{3.55}$$

Thus, the function estimate comprises a linear combination of $n + 1$ known functions: The first function is a constant, and for every measurement there corresponds a function $\gamma(\|\mathbf{x} - \mathbf{x}_i\|)$. This formulation has computational advantages. Once the ξ coefficients are computed, the estimate at any point can be computed with n multiplications. This formulation proves useful in graphing functions.

This formulation is also valuable because it reveals immediately properties of the best estimate function. For example:

- If the variogram is discontinuous at the origin (nugget effect), the estimate function $\hat{z}\,(\mathbf{x})$ will be discontinuous at observation points.

- If the variogram has linear behavior at the origin, the estimate will have sharp corners at observation points.
- The more the observations, the more complex will be the function since it consists of more terms.
- Away from observations, the estimate will be a "simple" and smooth function. The estimate will be a more "complex and interesting" function near observation points.

Thus, the properties of the function that is interpolated from the observations through kriging depend pretty much on the properties of the variogram used. It is also interesting to view the variogram as a "base function" or "spline" in a linear interpolation scheme.

Analytical expressions for the mean square error and the estimation error covariance function can be found in reference [86]. In summary, the conditional covariance function (which is neither stationary nor isotropic) is given from the expression

$$R_c(\mathbf{x}, \mathbf{x}') = -\gamma(\|\mathbf{x} - \mathbf{x}'\|) - \sum_{i=1}^{n}\sum_{j=1}^{n} \gamma(\|\mathbf{x} - \mathbf{x}_i\|) P_{ij} \gamma(\|\mathbf{x}_j - \mathbf{x}'\|)$$
$$- B + \sum_{j=1}^{n} a_j (\gamma(\|\mathbf{x} - \mathbf{x}_j\|) + \gamma(\|\mathbf{x}_j - \mathbf{x}'\|)), \tag{3.56}$$

where $\mathbf{P}$ is n by n, $\mathbf{a}$ is n by 1, and B is a scalar, obtained through the inversion of the matrix of the coefficients in system (3.55):

$$\begin{bmatrix} \mathbf{Q} & \mathbf{X} \\ \mathbf{X}^T & 0 \end{bmatrix}^{-1} = \begin{bmatrix} \mathbf{P} & \mathbf{a} \\ \mathbf{a}^T & B \end{bmatrix}, \tag{3.57}$$

where $\mathbf{Q}$ is an n by n matrix, $\mathbf{X}$ is an n by 1 vector, and $Q_{ij} = -\gamma(\|\mathbf{x}_i - \mathbf{x}_j\|)$, $X_i = 1$.

Exercise 3.6 *Consider data $z(x_1), \ldots, z(x_n)$ in a one-dimensional domain. Predict the form of function $\hat{z}(x)$ based on formula (3.54) without carrying out the computations for the following two cases:*

(a) nugget-effect variogram
(b) linear variogram.

Exercise 3.7 *Which of the basic variogram models presented in this chapter yields an estimate that is differentiable at any order?*

Exercise 3.8 *Give an intuitive interpretation to the interpolation method of Equation (3.54).*

3.11 Conditional realizations

In applications we are ultimately asking the question:

What is the actual function $z(\mathbf{x})$?

Due to inadequacy of information, there are many possible answers. It is often useful to generate a large number of equally likely (given the available information) possible solutions in order to perform probabilistic risk analysis. We call these solutions *conditional realizations,* but they are also known as conditional sample functions or conditional simulations. The average of the ensemble of all conditional realizations is equal to the best estimate and the covariance is equal to the mean square error matrix.

Conceptually, we start with the ensemble of unconditional realizations, such as those shown in Figures 3.2, 3.3, and 3.4. The ensemble-average variogram is the same as the model variogram,

$$\frac{1}{N}\sum_{k=1}^{N}\frac{1}{2}(z(\mathbf{x};k)-z(\mathbf{x}';k))^2 \rightarrow \gamma(\|\mathbf{x}-\mathbf{x}'\|), \tag{3.58}$$

as N tends to infinity. Then, from that ensemble, we keep only those realizations that are consistent or "honor" the observations $z(\mathbf{x}_1), \ldots, z(\mathbf{x}_n)$ to form the ensemble of conditional realizations. Thus, conditional realizations are consistent with both the structure and the specific observations.

We will provide a rudimentary description of how to generate a conditional realization, with emphasis on the key ideas and without worrying about optimizing algorithms.

3.11.1 Point simulation

Step 1 Generate an unconditional realization $z(x;k)$, with zero mean. This is a technical issue that goes beyond the scope of this book, but an introduction is given in Appendix C. (The index k is used as a reminder that there are many realizations.)

Step 2 Generate a conditional realization $z_c(\mathbf{x}_0;k)$ at a given point $\mathbf{x}_0$, through "rectification" of $z_c(\mathbf{x}_0;k)$:

$$z_c(\mathbf{x}_0;k) = z(\mathbf{x}_0;k) + \sum_{i=1}^{n}\lambda_i(z(\mathbf{x}_i) - z(\mathbf{x}_i;k)), \tag{3.59}$$

where $z(\mathbf{x}_i)$ is the actual observation and the λ coefficients are the same ones that were used to obtain the best estimate.

***Exercise 3.9** Prove that the expected value and variance of $z_c(\mathbf{x}_0; k)$ produced from the above procedure is the same as the best estimate and mean square error found from kriging.*

3.11.2 *Function simulation*

The functional form of $z_c(\mathbf{x}; k)$ is developed similarly to that of the best estimate in Section 3.10. First, generate an unconditional realization, a function $z(\mathbf{x}; k)$, with zero mean. Then

$$z_c(\mathbf{x}; k) = -\sum_{j=1}^{n} \gamma(\|\mathbf{x} - \mathbf{x}_j\|)\xi_j + \hat{\beta} + z(\mathbf{x}; k), \tag{3.60}$$

where the ξ and $\hat{\beta}$ coefficients are found from system

$$-\sum_{j=1}^{n} \gamma(\|\mathbf{x}_i - \mathbf{x}_j\|)\xi_j + \hat{\beta} = z(\mathbf{x}_i) - z(\mathbf{x}_i; k), \quad \text{for } i = 1, \ldots, n$$
$$\sum_{j=1}^{n} \xi_j = 0. \tag{3.61}$$

3.12 Properties of kriging

Kriging, or best linear unbiased estimation (BLUE) given only the variogram, has found many applications in mining, geology, and hydrology. It shares with other variants of BLUE techniques the following features: 1. The estimator is a linear function of the data with weights calculated according to the specifications of unbiasedness and minimum variance. Unbiasedness means that on the average the error of estimation is zero. Minimum variance means that, again on the average, the square estimation error is as small as possible. 2. The weights are determined by solving a system of linear equations with coefficients that depend only on the variogram that describes the structure of a family of functions. (Neither the function mean nor the variance are needed.)

In selecting the weights of the linear estimator, kriging accounts for the relative distance of measurements from each other and from the location where an estimate is sought. Consider, for example, the case in which z is a gradually varying function. Then, measurements in the neighborhood of the unknown value are given more weight than measurements located at a distance. Furthermore, kriging accounts for the fact that two measurements located near each other contribute the same type of information. Thus, the area of influence of each measurement is essentially taken into account.

In the case of interpolation, kriging is an "exact interpolator." That is, the contour surface of the estimate reproduces the measurements. In the case of variograms with a nugget (discontinuous at the origin), the contour map has a discontinuity at each observation point.

How does kriging compare to other methods used in interpolation and spatial averaging, such as inverse-distance weighing and deterministic splines, which are used in interpolation and contouring, or Thiessen polygons, a method used in hydrology for estimation of mean areal precipitation from point measurements [90]? A major advantage of kriging is that it is more flexible than other interpolation methods. The weights are not selected on the basis of some arbitrary rule that may be applicable in some cases but not in others, but depend on how the function varies in space. Data can be analyzed in a systematic and objective way, as we will see in the following chapters, and prior experience is used to derive a variogram that is then used to determine the appropriate weights. Depending on the scale of variability, we may use equal or highly variable weights, whereas Thiessen polygons, to mention one example, applies the same weights, whether the function exhibits small- or large-scale variability.

Another advantage of kriging is that it provides the means to evaluate the magnitude of the estimation error. The mean square error is a useful rational measure of the reliability of the estimate; it depends only on the variogram and the location of the measurements. Thus, given the variogram, one can evaluate the mean square error of an estimate for a *proposed* set of measurement locations. A useful application of kriging is in the design of sampling networks or in the selection of the location of the next measurement, as in references [7, 71, and 120].

An inherent limitation of the linear estimation methods discussed in this chapter is that they implicitly assume that the available information about the spatial structure of the spatial function can be described adequately through a variogram. Although this is often an acceptable assumption, one must keep in mind that distinctly different types of functions may have the same variogram. Intuitively, this means that if we only specify the variogram, we specify a family of functions that contains many different solutions. If additional information is available about the shape of the spatial function allowing us to eliminate many candidates, then the kriging method of this chapter may not be the best approach because it cannot directly use this information. Other estimation methods have been developed that extend the BLUE methodology, and we will see some of these methods. Also, methods outside the BLUE family have been developed, including disjunctive kriging, probability kriging, and indicator kriging [37, 75, 98, 114, 115, 143, 112, 117, 132]; however, such methods are beyond the scope of this book.

3.13 Uniqueness

The reader may wish to skip this section at first reading.

Will the kriging system always have a unique solution? The answer is yes, provided that we have used a mathematically acceptable variogram and there are no redundant measurements. If we superpose variograms from the approved list with parameters in the right range, the variogram we obtain is mathematically acceptable.

Let us clarify what we mean by redundant measurements. Consider two measurements obtained at the same location. If the variogram is continuous (no nugget effect), then the function z is continuous, which means that one of the two measurements is redundant (*i.e.*, it contains no information that is not already given by the other measurement). One of the two measurements must be discarded; otherwise, a unique solution cannot be obtained because the determinant of the matrix of coefficients of the kriging system vanishes. In practice, the same conclusion holds when two measurements are located at a very short distance h_s and $\gamma(h_s)$ is very near zero at this short distance, as sometimes happens when using the Gaussian variogram. Any attempt to solve this system on a digital computer will encounter numerical difficulties.

There are only two possible solutions to this problem: Either keep the model but get rid of the redundant measurement or modify the model to make the two measurements necessary, thereby removing the numerical problem. This modification can be accomplished by adding a nugget term to the variogram.

Note that, when the variogram contains a nugget, two observations at the same location or at nearby locations may differ significantly, as the result of random measurement error and microvariability. In such a case, it may be desirable to keep both measurements. The value of $\gamma(\mathbf{x}_i - \mathbf{x}_j)$ should be taken equal to C_0, the nugget-effect variance, even though $\mathbf{x}_i \simeq \mathbf{x}_j$. The two measurements may be interpreted as being taken at a very small but nonzero separation distance from one another.

For readers interested in a more rigorous mathematical treatment of the issue of uniqueness of the solution, necessary and sufficient conditions can be found at page 226 in reference [92].

3.14 Kriging as exact interpolator

You may choose to skip this section at first reading.

Consider the case in which the location of the unknown coincides with the location of the i-th measurement, *i.e.*, $x_0 = x_i$. For example, consider kriging with two measurements and $x_0 = x_1$. The kriging system consists of three equations:

$$\begin{array}{rcrcrcl} 0\lambda_1 & - & \gamma(x_1 - x_2)\lambda_2 & + & \nu & = & -\gamma(x_0 - x_1) \\ -\gamma(x_2 - x_1)\lambda_1 & + & 0\lambda_2 & + & \nu & = & -\gamma(x_0 - x_2) \\ \lambda_1 & + & \lambda_2 & + & 0\nu & = & 1. \end{array} \qquad (3.62)$$

Let us distinguish between the cases of continuous and discontinuous (nugget-effect) functions.

1. For continuous functions (*e.g.*, $\gamma(x - x') = |x - x'|$), set $\gamma(x_i - x_0) = 0$ on the right-hand side of the kriging system. The solution is $\lambda_1 = 1$, $\lambda_2 = 0$, and $\nu = 0$. Kriging, therefore, reproduces the measured value, and the mean square estimation error, calculated from (3.40), is zero.
   Of course, the same would be true for any number of measurements. Thus, in the case of continuous functions kriging is an "exact" interpolator. The estimate $\hat{z}(x)$ varies in a continuous fashion and can be plotted.
2. What about the case in which $x_0 = x_i$, but with a variogram that exhibits a nugget effect? Setting $\gamma(x_i - x_0) = 0$ on the right-hand side of the kriging system does guarantee that $\hat{z}(x_0) = z(x_i)$. Thus, it is mathematically straightforward to make kriging an exact interpolator if it is so desired. However, the practical significance or usefulness of such an exact interpolator can be questioned in this case. Let us use the example with the two measurements and assume that $\gamma(x - x') = C_0$, for $|x - x'| > 0$, while, as always, $\gamma(0) = 0$. When $x_0 = x_1$, one can verify that $\lambda_1 = 1$, $\lambda_2 = 0$, and $\nu = 0$. Thus, the kriging equations we have developed reproduce the measured value, and the mean square estimation error is zero. However, for any other x_0, no matter how close it is to x_1, $\lambda_1 = 1/2$, $\lambda_2 = 1/2$, and $\nu = -C_0/2$. Kriging uses the arithmetic average of the measurements, and the mean square estimation error is computed to be $3C_0/2$. As a result, the estimate $\hat{z}(x)$ is a *discontinous* function.

Exercise 3.10 *What is the meaning of the nugget effect? How would you explain it in nontechnical terms? What are the practical consequences of introducing or changing the intensity of the nugget effect in interpolation using kriging? (Hint: You will find it useful to compare the extreme cases of no nugget versus the pure nugget effect. You can experiment with a kriging program.)*

3.15 Generalized covariance functions

As with the last two sections, you may wish to initially bypass this section.

It is rather straightforward to express the kriging equations in terms of covariance functions. For stationary functions with unknown mean, the kriging coefficients and the mean square estimation error are given below:

$$\sum_{j=1}^{n} \lambda_j R(\mathbf{x}_i - \mathbf{x}_j) + \nu = R(\mathbf{x}_i - \mathbf{x}_0) \quad i = 1, 2, \ldots, n$$

$$\sum_{j=1}^{n} \lambda_j = 1 \tag{3.63}$$

$$E[(\hat{z}_0 - z(\mathbf{x}_0))^2] = -\nu - \sum_{i=1}^{n} \lambda_j \lambda_i R(\mathbf{x}_i - \mathbf{x}_0) + R(0).$$

These equations apply when the covariance function can be defined. This is a good time to introduce the concept of the *generalized covariance* function. The motivation is that it is so much more convenient to work with covariance functions than with variograms. For the intrinsic case, the generalized covariance function, $K(h)$, is

$$K(h) = -\gamma(h) + C, \tag{3.64}$$

where C is an arbitrary constant. Note that if we use $K(h)$ in the place of $R(h)$ in Equations (3.63)–(3.63), the kriging coefficients or the mean square error are not affected by the value of C. Alternatively, we may use $K(h)$ instead of $-\gamma(h)$ in Equations (3.41) and (3.48). The usefulness of the generalized covariance function will be appreciated when we work with more complicated models, such as with a trending mean. We will stick with the variogram in this chapter.

Example 3.3 *(continuation of Example 1): In this example the variance is infinitely large. Nevertheless, we can use the generalized covariance function*

$$K(h) = \begin{cases} -h, & \text{if } h > 0 \\ 1, & \text{if } h = 0 \end{cases}.$$

The kriging system of equations needed to estimate the value at $x_0 = 2$ *is*

$$\begin{bmatrix} 1 & -1 & -3 & 1 \\ -1 & 1 & -2 & 1 \\ -3 & -2 & 1 & 1 \\ 1 & 1 & 1 & 0 \end{bmatrix} \begin{bmatrix} \lambda_1 \\ \lambda_2 \\ \lambda_3 \\ \nu \end{bmatrix} = \begin{bmatrix} -2 \\ -1 \\ -1 \\ 1 \end{bmatrix}.$$

The mean square estimation error is $-\nu + 2\lambda_1 + \lambda_2 + \lambda_3 + 1$.

Solving the system, we obtain $\lambda_1 = 0.1304$, $\lambda_2 = 0.3913$, $\lambda_3 = 0.4783$, and $\nu = -0.304$. The mean square estimation error is $MSE = 2.43$. That is, the solution is the same whether we use variograms or (generalized) covariance functions.

3.16 Key points of Chapter 3

Typically, the problem involves finding a spatially variable quantity, $z(\mathbf{x})$, from a few measurements, $z(\mathbf{x}_1), \ldots, z(\mathbf{x}_n)$. We perform the search in two phases:

1. structural analysis, where we determine the family of functions that includes the solution and
2. conditioning, where we eliminate from consideration members of the family that are inconsistent with the data.

In structural analysis, instead of studying the multitude of all possible solutions individually, we deal with their averages. In practice, we work with the first two statistical moments and so structural analysis entails fitting equations that describe the first two moments. The intrinsic model is the simplest model and basically postulates that the mean is constant (with a value that is not needed) and the mean square difference is a function of the separation distance only. This function, called the variogram, is a way to describe the distribution of variability among scales. Once the variogram is chosen, one can perform conditioning through best linear unbiased estimation. In this method, the estimate is a linear function (in a sense, a "weighted average") of the data with coefficients that are selected to minimize the mean square error and to render an expected error of zero.

4
Variogram fitting

To apply the most common geostatistical model in the solution of an interpolation problem, the variogram must be selected. In this chapter, we discuss the practice of

1. fitting a variogram to the data, assuming that the intrinsic model is applicable, and
2. checking whether the intrinsic model is in reasonable agreement with the data.

4.1 The problem

Let $z(\mathbf{x})$ be the spatial function that needs to be estimated, such as concentration as a function of the location $\mathbf{x}$ in an aquifer. Consider that n measurements have been collected:

$$z(\mathbf{x}_1),\ z(\mathbf{x}_2), \ldots, z(\mathbf{x}_n).$$

We need to fit a model that can be used for interpolation of $z(\mathbf{x})$ from the measurement points to other locations. In Chapter 3, we saw that the intrinsic model can be used for the solution of interpolation problems. For any location $\mathbf{x}_0$, using a method called kriging we can obtain the best estimate of $z(x_0)$ as well as the mean square error associated with this estimate. In other words, we can obtain the most accurate estimate and a measure of how close this estimate is to the true value. To apply this approach to actual data, however, we have to

1. choose the variogram, assuming that the intrinsic model is applicable, and
2. ascertain whether the intrinsic model is appropriate.

These questions are related. However, for instructional purposes, they will be treated separately. We will start with the first question. Unless otherwise mentioned, in this chapter it will be taken for granted that the intrinsic (and isotropic) model is applicable and the focus will be on determining the variogram that fits the data as closely as possible. Finally, after an appropriate variogram has been selected, we will deal with the second question.

4.2 Prior information

Although there is no disagreement about the practical usefulness of the variogram, there are diverse views on what it really represents. Two extreme positions are:

- The variogram is not dependent on the data but rather is introduced based on other information.
- The variogram is estimated only from the data.

We will discuss some of these issues. Let us start by considering the statement "even if no observations of field variable z were available, its variogram could still be devised based on other information." Of course, this statement makes no sense if we consider the field variable z in abstract terms. However, geostatistical estimation methods should be examined in the context of their application, not as abstract theory. Thus, $z(x)$ to hydrogeologist A may be transmissivity in a specific aquifer, to environmental engineer B concentration of given pollutant, and so on. Once we consider what we know about a variable, we realize that some of this information can be translated into a geostatistical model.

If we had absolutely no knowledge of $z(\mathbf{x})$, the nugget-effect variogram with infinite variance would be appropriate. Kriging would yield estimates with infinite mean square error at unmeasured locations, which is reasonable since the adopted variogram means that z could be anything. However, in practice, this situation is definitely not realistic. For example, if z represents log-transmissivity, its variance is unlikely to be larger that 10 and is probably less than 4; also, the transmissivity function is likely to exhibit some degree of spatial continuity (see reference [66]). When these considerations are taken into account, we may decide to adopt an exponential variogram with variance and length parameters selected based on our judgement and experience. If this variogram is then used for kriging, the mean square error of estimation represents a rational measure of prediction reliability that reflects our judgement and experience.

On the other extreme, conventional statistical methodology is concerned with making statements based on data and involves as little subjective judgement as possible. Such methods can be used to infer some features of the variogram with accuracy that tends to improve as more observations are collected.

In practice, the selection of the variogram (or, more generally, the geostatistical model) should follow two principles:

Consistency with observations The variogram should be in reasonable agreement with the data. Thus, the variogram depends on the observations, particularly if the data set is large.

Consistency with other information Certain features of the variogram cannot be determined from the data and may be resolved based on other information. Thus, the variogram also depends on the analyst's judgement and experience.

In what follows, we focus on developing variograms that are consistent with the data or, in other words, on how to extract from data information about the variogram.

4.3 Experimental variogram

Variogram selection is an iterative process that usually starts with an examination of the experimental variogram. We already have discussed how we can obtain the experimental variogram from the data. Then, we fit a variogram model by selecting one of the equations from the list of basic models of the previous chapter and adjusting its parameters to reproduce the experimental variogram as closely as possible. For example, we may select the linear variogram, $\gamma(h) = \theta h$, and then fit the parameter θ to trace the experimental variogram, as illustrated in Figure 4.1a. If we cannot obtain a reasonably good fit with a single basic model, we may superpose two or more of them. For example, superposing the linear plus nugget-effect equations, we obtain the variogram model

$$\gamma(h) \begin{cases} C_0 + \theta h, & h > 0 \\ 0, & h = 0 \end{cases}, \tag{4.1}$$

which has two parameters that can be fitted to the data, $C_0 \geq 0$, $\theta \geq 0$, (see Figure 4.1b). Although one could superpose as many models as one might

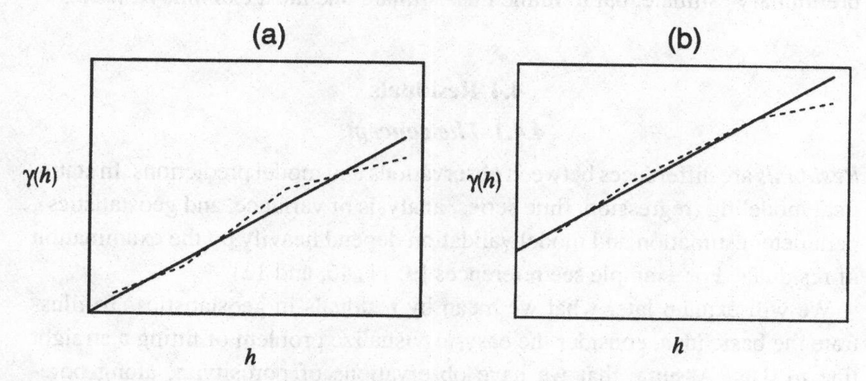

Figure 4.1 Variogram (*continuous line*) fitted to experimental variogram (*dashed line*).

want, in practice, one should use models with as few parameters as possible. Furthermore, one should avoid superposing two models that have very similar shapes (such as the exponential and the spherical). We will discuss some of these issues later.

Fitting an equation to the experimental variogram enables us to get started with the analysis and to obtain a reasonable rough estimate of the variogram. In many studies one needs to refine this estimate. However, the experimental variogram cannot provide a solid basis for model refinement and testing, for a number of reasons:

1. The i-th node of the experimental variogram, $\hat{\gamma}(h_i)$, is only an *estimate* of the value of $\gamma(h_i)$. Associated with this estimate is a measure of reliability, which depends on many factors and may be quite different from one node to the next. Thus, it is not clear how one should fit the equation to the experimental variogram. For example, in most cases, it is desired to obtain a better fit near the origin because the reliability of the experimental variogram nodes decreases at large separation distance.
2. The experimental variogram is often sensitive to the discretization of the separation distance into intervals. There is no universally satisfactory way to select intervals because using longer intervals means that you trade resolution for reliability.
3. Fitting the model to the experimental variogram is not necessarily the same as fitting the model to the original data. The degree of agreement between the model and the experimental variogram may have nothing to do with the adequacy of the model or the accuracy with which the model can reproduce the data.

These points will become clearer when we see applications. For now, keep in mind that fitting an equation to the experimental variogram usually gives a good preliminary estimate, but to refine this estimate one must examine *residuals*.

4.4 Residuals

4.4.1 The concept

Residuals are differences between observations and model predictions. In statistical modeling (regression, time series, analysis of variance, and geostatistics), parameter estimation and model validation depend heavily on the examination of residuals. For example see references [9, 11, 46, and 12].

We will explain later what we mean by residuals in geostatistics. To illustrate the basic idea, consider the easy-to-visualize problem of fitting a straight line to data. Assume that we have observations of porosity, z, along bore-

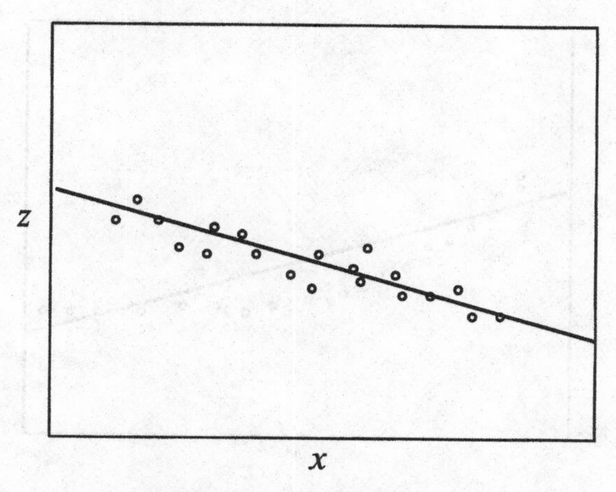

Figure 4.2 Fitting a straight line to data.

hole depth, x. The porosity seems to decrease with depth and a straight line (the "model") is chosen to represent spatial variability. Then, the residuals are $z_i - (ax_i + b)$, where z_i is observation at point i and $ax_i + b$ represents the model prediction at location x_i; a and b are parameters. See Figure 4.2, which depicts the data as points and the model prediction with a continuous line. From this example, we can see that residuals can be useful for two purposes:

1. To adjust the parameters of the model to get the best fit. A common approach is to select the coefficients a and b that minimize the sum of squares of the residuals,

$$\sum_i [z_i - (ax_i + b)]^2. \tag{4.2}$$

2. To evaluate whether another model, such as a quadratic function, may be a "better" model. If the residuals exhibit a nonrandom pattern, as illustrated in Figure 4.3, then another model that accounts for this pattern may represent the data better than the present model. (Whether an alternative model is appropriate is a different issue. The residuals simply point out patterns missed by the model currently being tested.) The idea is that we view the porosity data as the sum of a "predictable" part, which we try to capture with our model, and a "random" part. The residuals should represent the random part.

Thus, residuals prove useful in parameter estimation (or model calibration) and in model testing. In linear estimation, such as the kriging method of Chapter 3, residuals play as important a role as in the example above.

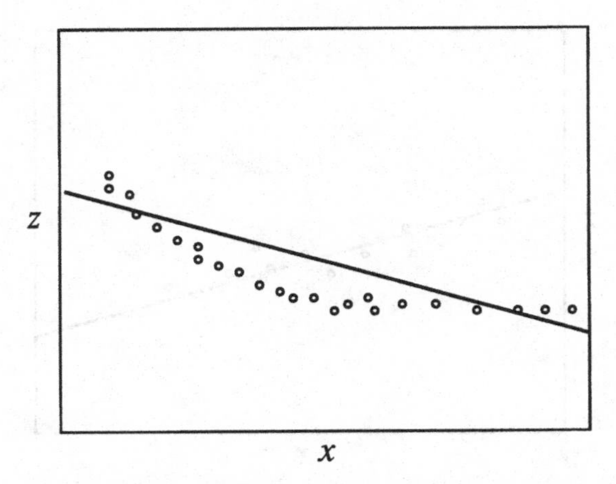

Figure 4.3 Fitting a straight line to data that would better fit a quadratic.

4.4.2 Residuals in kriging

Now, let us see what we mean by residuals in the case of kriging, following reference [82].

The model is one where the unknown spatial function $z(\mathbf{x})$ is a realization of an intrinsic function with variogram $\gamma(h)$. We will pretend that the n measurements are given to us one at a time, in a given sequence. The sequential processing is an important part of the methodology. (For now, we do not need to concern ourselves with how to arrange the measurements in a sequence. For many applications, any sequence will give the same result.)

Next, calculate the kriging estimate of z at the second point, $\mathbf{x}_2$, given only the first measurement, $\mathbf{x}_1$. Obviously, $\hat{z}_2 = z(\mathbf{x}_1)$ and $\sigma_2^2 = 2\gamma(\mathbf{x}_1 - \mathbf{x}_2)$. Calculate the actual error,

$$\delta_2 = z(\mathbf{x}_2) - \hat{z}_2, \tag{4.3}$$

and normalize by the standard error

$$\varepsilon_2 = \frac{\delta_2}{\sigma_2}. \tag{4.4}$$

Use the same procedure to construct the other residuals. For the k-th measurement location, estimate through kriging the value of z using only the first $k-1$ measurements and normalize by the standard error. Thus,

$$\delta_k = z(\mathbf{x}_k) - \hat{z}_k, \quad \text{for } k = 2, \ldots, n \tag{4.5}$$

$$\varepsilon_k = \frac{\delta_k}{\sigma_k}, \quad \text{for } k = 2, \ldots, n. \tag{4.6}$$

Note that the ε residuals are a normalized version of the δ residuals and that we end up with $n-1$ residuals, δ or ε.

4.4.3 Distribution of residuals

Using the actual data, the variogram model, and kriging, one can compute the actual (or *experimental*) residuals. It is useful (for example, in model validation) to invoke the concept of the stochastic process to establish a probability distribution for the residuals.

Consider the ensemble of possible realizations of a random function $z(\mathbf{x})$. Using the observation locations, the variogram, and the method of kriging, we can in theory compute the residuals for each realization. This is the way to generate the ensemble of all possible realizations of the residuals. In this sense, the residuals are random variables and we can compute their ensemble statistics (mean values, variances, covariances, etc.) or probability distributions.

If the model is correct (*i.e.*, that the process is intrinsic with the given variogram), the residuals satisfy the following important relations:

$$E[\varepsilon_k] = 0, \quad k = 2, \ldots, n \tag{4.7}$$

$$E[\varepsilon_k \varepsilon_l] = \begin{cases} 1, & \text{if } k = \ell \\ 0, & \text{if } k \neq \ell \end{cases}, \quad k, \ell = 2, \ldots, n. \tag{4.8}$$

Proof That $E[\varepsilon_k] = 0$ follows from the unbiasedness property of kriging. That $E[\varepsilon_k^2] = 1$ is a consequence of the normalization. Now assume that $\ell > k$ and, to simplify the algebra, assume that a covariance function R has been defined. Then

$$\begin{aligned} E[\varepsilon_k \varepsilon_l] &= \frac{1}{\sigma_k \sigma_l} E\left[\left(z(\mathbf{x}_k) - \sum_{i=1}^{k-1} \lambda_{ki} z(\mathbf{x}_i)\right)\left(z(\mathbf{x}_\ell) - \sum_{j=1}^{\ell-1} \lambda_{lj} z(\mathbf{x}_j)\right)\right] \\ &= \frac{1}{\sigma_k \sigma_\ell}\left\{\left(R(\mathbf{x}_k - \mathbf{x}_\ell) - \sum_{j=1}^{\ell-1} \lambda_{\ell j} R(\mathbf{x}_k - \mathbf{x}_j)\right)\right. \\ &\quad \left. - \sum_{i=1}^{k-1} \lambda_{ki}\left[R(\mathbf{x}_i - \mathbf{x}_\ell) - \sum_{j=1}^{\ell-1} \lambda_{\ell j} R(\mathbf{x}_i - \mathbf{x}_j)\right]\right\}, \end{aligned} \tag{4.9}$$

where λ_{ki} and $\lambda_{\ell j}$ are kriging coefficients determined from the kriging system; for any $i, i < \ell$, the following relation holds:

$$\sum_{j=1}^{\ell-1} \lambda_{\ell j} R(\mathbf{x}_i - \mathbf{x}_j) + \nu_\ell = R(\mathbf{x}_i - \mathbf{x}_\ell), \quad i = 1, 2, \ldots, \ell - 1. \tag{4.10}$$

Consequently,

$$E[\varepsilon_k \varepsilon_\ell] = \frac{1}{\sigma_k \sigma_\ell}\left(\nu_\ell - \sum_{i=1}^{k-1} \lambda_{ki} \nu_\ell\right) = 0. \tag{4.11}$$

We say that the εs are *orthonormal* (*i.e.*, uncorrelated from each other and normalized to have unit variance). In optimal filtering, such sequences are known as *innovations* (meaning "new information") since each ε_k contains information about $z(\mathbf{x}_k)$ that could not be predicted from the previous measurements $z(\mathbf{x}_1), \ldots, z(\mathbf{x}_{k-1})$. From

$$z(\mathbf{x}_k) = \hat{z}_k + \sigma_k \varepsilon_k \tag{4.12}$$

we can see that the measurement at $\mathbf{x}_k$ is the sum of the best estimate using previous measurements (the predictable part) and the innovation. The lack of correlation in the orthonormal residuals can be explained in intuitive terms as follows: If the sequence of the residuals were correlated, one could use this correlation to predict the value of ε_k from the value of $\varepsilon_2, \ldots, \varepsilon_{k-1}$ using a linear estimator. This allows one to reduce further the mean square error of estimation of $z(\mathbf{x}_k)$; but this is impossible because $\hat{z}_k$ is already the minimum-variance estimate. Thus, the orthonormal residuals must be uncorrelated if the model is correct and kriging is a minimum variance unbiased estimator.

Exercise 4.1 *You are given measurements $z(\mathbf{x}_1), \ldots, z(\mathbf{x}_n)$. The model is an intrinsic function with a pure-nugget effect variogram:*

$$\gamma(\|\mathbf{x} - \mathbf{x}'\|) = \begin{cases} C_0, & \text{if } \mathbf{x} \neq \mathbf{x}' \\ 0, & \text{if } \mathbf{x} = \mathbf{x}' \end{cases}.$$

Show that the residuals are

$$\delta_k = z(\mathbf{x}_k) - \frac{1}{k-1} \sum_{i=1}^{k-1} z(\mathbf{x}_i), \quad \text{for } k = 2, \ldots, n$$

and

$$\varepsilon_k = \frac{\delta_k}{\sqrt{\frac{kC_0}{k-1}}}, \quad \text{for } k = 2, \ldots, n.$$

4.5 Model validation

Model validation means testing the model. Every empirical model must be tested before it is used for predictions. The confidence we have in model predictions relies to a great extent on how the model has performed in tests that have the potential to discredit the model.

In practice, model validation is based on statistical tests involving the residuals. A statistical test is the equivalent of an experiment that is conducted to validate a scientific theory. Assume that a theory ("model" or *null hypothesis*, denoted by H_0) has been proposed and we want to validate it. We design an experiment and then we

1. predict the outcome to the experiment using the theory;
2. observe the actual outcome of the experiment; and
3. compare the anticipated and observed outcomes.

If the agreement is acceptable, such as when the difference is within the anticipated error, then we say that the data provide no reason to reject the model. Otherwise, the data discredit the theory, *i.e.*, give us reason to reject the model.

We follow the same approach in statistical estimation of spatial functions. Some examples will illustrate the idea.

4.5.1 Q_1 statistic

Experiment Compute the orthonormal residuals, as described in Section 4.4.2, then compute their average:

$$Q_1 = \frac{1}{n-1} \sum_{k=2}^{n} \varepsilon_k. \tag{4.13}$$

Model Prediction Our model involves a family (or ensemble) of functions, as we mentioned at the beginning of Chapter 3, with variogram $\gamma(h)$. (If this is unclear, reread Section 3.1.) For each function with probability of occurrence P_i, we compute a value of Q_1 with probability of occurrence P_i. That is, our theory says that Q_1 is a random variable and allows us to compute its probability distribution. We start with its mean value

$$E[Q_1] = E\left[\frac{1}{n-1}\sum_{k=2}^{n} \varepsilon_k\right] = \frac{1}{n-1}\sum_{k=2}^{n} E[\varepsilon_k] = 0, \tag{4.14}$$

followed by the variance

$$\begin{aligned} E\left[Q_1^2\right] &= E\left[\left(\frac{1}{n-1}\sum_{k=2}^{n} \varepsilon_k\right)^2\right] = E\left[\left(\frac{1}{n-1}\right)^2 \sum_{k=2}^{n}\sum_{l=2}^{n} \varepsilon_k\, \varepsilon_l\right] \\ &= \left(\frac{1}{n-1}\right)^2 \sum_{k=2}^{n}\sum_{l=2}^{n} E[\varepsilon_k\, \varepsilon_l] = \left(\frac{1}{n-1}\right)^2 (n-1) = \frac{1}{n-1}. \end{aligned} \tag{4.15}$$

The distribution of Q_1 is considered normal for the following reasons: (*a*) Implicit in our linear estimation is that the residuals are approximately normal.

The distribution of a random variable that is a linear function of normal random variables is normal. (*b*) Even if the orthonormal residuals are not exactly normal, the distribution of their average will tend more and more toward the normal one as the number of terms included increases, as postulated by the *central limit theorem* of probability theory.

Thus, under the null hypothesis, Q_1 is normally distributed with mean 0 and variance $\frac{1}{n-1}$. The pdf of Q_1 is

$$f(Q_1) = \frac{1}{\sqrt{2\pi/(n-1)}} \exp\left(-\frac{x^2}{2/(n-1)}\right). \tag{4.16}$$

The important point is that we expect that all sample values of Q_1 must be near zero, especially if n is large. For example, there is only about a 5% chance that a Q_1 value will be larger than $2/\sqrt{n-1}$ in absolute value. See Figure 4.4, which shows the pdf of Q_1 for $n = 20$ as well as the fences at $\pm 2/\sqrt{n-1} = \pm 0.46$. The area under the curve between these two fences is approximately 0.95.

Experimental Observation Measure the experimental value of Q_1 (the numerical value from the actual data).

Compare Model with Experiment If the experimental value of Q_1 turns out to be pretty close to zero then this test gives us no reason to question the validity of the model. However, if it is too different from zero, then the model is discredited. We may adopt the following:

Decision rule Reject the model if

$$|Q_1| > \frac{2}{\sqrt{n-1}}. \tag{4.17}$$

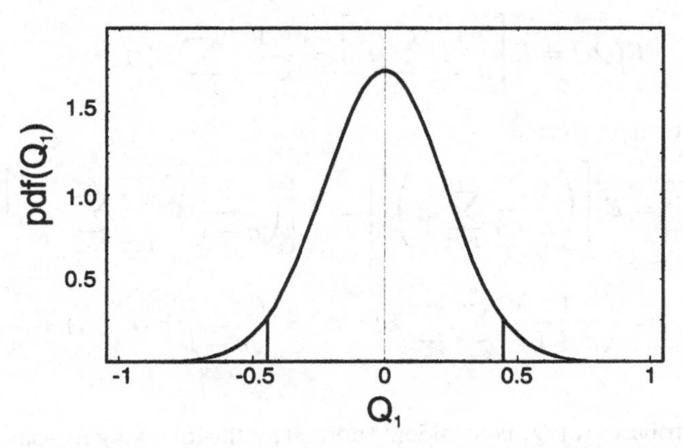

Figure 4.4 Probability density function of Q_1 statistic for $n = 20$.

This rule involves a 5% probability that the correct model may be rejected. We may reduce this risk by increasing the value of the threshold $(2/\sqrt{n-1})$, but by doing so we reduce the chance to weed out a bad model. The 5% cutoff, although customary in statistics, is somewhat arbitrary. One may adjust the test on the basis of the prior confidence in the model. That is, if one has reasons to trust the model, then one may use $2.5/\sqrt{n-1}$. The bottom line, however, is that a large $|Q_1|$ value is a sign that the model does not fit the data.

4.5.2 Q_2 statistic

The same procedure can be applied with several other statistics. The most useful one is

$$Q_2 = \frac{1}{n-1} \sum_{k=2}^{n} \varepsilon_k^2. \tag{4.18}$$

Again, under the null hypothesis, Q_2 is a random variable with mean

$$E[Q_2] = 1 \tag{4.19}$$

and variance that can be calculated for normally distributed z,

$$E[(Q_2 - 1)^2] = \frac{2}{n-1}. \tag{4.20}$$

Furthermore, under the assumption that the residuals are approximately normal, $(n-1)Q_2$ follows the chi-square distribution with parameter $(n-1)$. The pdf of Q_2 can then be derived:

$$f(Q_2) = \frac{(n-1)^{\frac{n-1}{2}} Q_2^{\frac{n-3}{2}} \exp\left(-\frac{(n-1)Q_2}{2}\right)}{2^{\frac{n-1}{2}} \Gamma\left(\frac{n-1}{2}\right)}, \tag{4.21}$$

where Γ is the gamma function. A plot of this distribution for $n = 20$ is shown in Figure 4.5. The key point is that most values should be near 1. Specifically, for $n = 20$, there is a probability 0.95 that the value of Q_2 is between 0.47 and 1.73.

Thus, we can come up with another 5% rule:

Decision rule Reject the model if

$$Q_2 > U \quad \text{or} \quad Q_2 < L, \tag{4.22}$$

where the values of U and L are known from the pdf of $f(Q_2)$ (see Table 4.1, which was adapted from a table of the χ^2 distribution in [62].) Actually, for

Table 4.1. *The 0.025 and 0.975 percentiles of the Q_2 distribution*

$n-1$	L	U	$n-1$	L	U
1	0.001	5.02	21	0.490	1.69
2	0.025	3.69	22	0.500	1.67
3	0.072	3.12	23	0.509	1.66
4	0.121	2.78	24	0.517	1.64
5	0.166	2.56	25	0.524	1.62
6	0.207	2.40	26	0.531	1.61
7	0.241	2.29	27	0.541	1.60
8	0.273	2.19	28	0.546	1.59
9	0.300	2.11	29	0.552	1.58
10	0.325	2.05	30	0.560	1.57
11	0.347	1.99	35	0.589	1.52
12	0.367	1.94	40	0.610	1.48
13	0.385	1.90	45	0.631	1.45
14	0.402	1.86	50	0.648	1.43
15	0.417	1.83	75	0.705	1.34
16	0.432	1.80	100	0.742	1.30
17	0.445	1.78			
18	0.457	1.75			
19	0.469	1.73			
20	0.479	1.71			

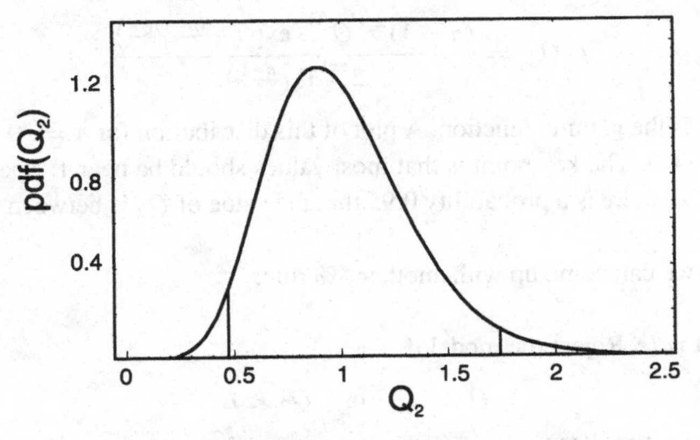

Figure 4.5 Probability density function of Q_2 for $n = 20$.

Table 4.2. *Coefficients for the Filliben test for departures from normality*

n	R	n	R
5	0.879	45	0.974
10	0.917	50	0.977
15	0.937	60	0.980
20	0.950	70	0.982
25	0.958	80	0.984
30	0.964	90	0.985
35	0.968	100	0.987
40	0.972		

$n > 40$, approximating the distribution of Q_2 as Gaussian is adequate for this application, so that one can use the following rule:

Reject the model if

$$|Q_2 - 1| > 2.8/\sqrt{n-1}. \tag{4.23}$$

4.5.3 Normality

One should also test whether the ε residuals follow a Gaussian distribution, because minimum-variance linear unbiased estimation methods implicitly assume that the estimation errors are approximately normal. Plotting $\varepsilon_2, \ldots, \varepsilon_n$ on normal probability paper allows one to detect visually departures from normality.

Another approach is to perform formal *goodness of fit* tests, such as the Shapiro-Wilk [126] and Filliben [52] tests. We do not intend to deal with all the details of such tests. In simple terms, we proceed as follows: We evaluate a number r between 0 and 1 that measures how close the probability distribution of the residuals is to the normal distribution with mean 0 and variance 1. The value of r varies from realization to realization, *i.e.*, r is a random variable with probability distribution that depends only on the value of n. We can compute the number R that is exceeded 95% of the time (see Table 4.2); in other words, the value of r should be between R and 1 in 95% of the times. For example, for $n = 21$, $R = 0.972$. If the experimental value is less than this value R, then we have reason to suspect that the residuals may not be normally distributed.

4.5.4 No correlation

Finally, if the model is correct, the ε residuals should be uncorrelated, *i.e.*, they should have the pure nugget-effect variogram,

$$\gamma(x_i - x_j) = \begin{cases} 1, & \text{for } x_i \neq x_j \\ 0, & \text{for } x_i = x_j \end{cases}. \tag{4.24}$$

One may thus perform a variogram analysis and test the hypothesis that the experimental variogram of $\varepsilon_2, \ldots, \varepsilon_n$ is indeed this γ. One can also plot the signs of the residuals seeking patterns indicating disagreement between the actual and the expected behavior of the model.

4.5.5 Ordering

As already mentioned, an important feature of the methodology is that the data are ordered. As long as no other model is being considered as a likely candidate, this order should be picked at random. However, if a reasonable alternate model exists, it makes sense to order the data to increase the probability that the original model will be rejected if the alternate model is valid. We will return to this issue after we consider models with variable mean.

Thus, we have seen that the orthonormal ε residuals are useful in testing the validity of the model. The idea is simple: If the model is good, we expect that the orthonormal residuals will have zero mean and unit variance and will be uncorrelated from each other. If the actual residuals do not behave this way, then the model can probably be improved.

4.6 Variogram fitting

In this section, we will see that the δ residuals are particularly important in evaluating how closely the model fits the data. The smaller the values of the δ residuals the better the fit. We can thus select parameters that optimize the fit.

Consider that we have tentatively accepted the model and selected an expression for the variogram; next, we must fit its parameters to the data. For example, we have chosen

$$\gamma(h) = v\left[1 - \exp\left(-\frac{h}{\ell}\right)\right] \tag{4.25}$$

but we need to fit the parameters v and ℓ to the data. In general terms, we evaluate the agreement between the model and the data by examining the differences between the data and the predicted values (the δ residuals). To optimize the agreement, we must select the parameters that minimize, in some sense, the residuals. Note that parameters impact the residuals in two ways: They affect the residuals δ_i (through their influence on the best estimates) and the estimation variances σ_i^2. For instance, in the variogram of Equation (4.25), δ_i depends only on parameter ℓ, whereas σ_i^2 is proportional to parameter v and is less sensitive to parameter ℓ.

To ensure that the model neither overestimates nor underestimates the variance, we select parameters so that

$$Q_2 = \frac{1}{n-1} \sum_{i=2}^{n} \frac{\delta_i^2}{\sigma_i^2} = 1. \tag{4.26}$$

If, for example, ℓ is known in Equation (4.25) and the only parameter to be estimated is v, then we select v so that $Q_2 = 1$. However, if both parameters need to be estimated, then this requirement does not specify a unique answer because there are many combinations of v and ℓ that satisfy (4.26). To choose among them, we must account for the size of the δ residuals. That is, we must select the values of the parameters that make the δ residuals as small as possible.

The underlying idea is simple: The right parameters should make the model reproduce the data. The challenge lies in finding the index that best represents in an average sense the size of the residuals; then, the best estimates of the parameters are the values that minimize the value of this index while at the same time satisfying (4.26).

A reasonable index is the mean square δ residual,

$$M = \frac{1}{n-1} \sum_{i=2}^{n} \delta_i^2. \tag{4.27}$$

An important advantage of this criterion is that it is simple and easy to explain. If the δ residuals follow the same (approximately normal distribution with zero mean), then M is the most representative measure of the size of the residuals. In such a case, the most reasonable estimates of the parameters are the ones that minimize the value of M.

However, in most applications the measurements are distributed in space in such a way that a few of the δ^2 values may determine the value of M. The largest values usually correspond to measurements located far away from other measurements and are consequently more erratically predicted from other measurements. Thus, a limitation of this method is that the parameter estimates are too sensitive to the values of the erratic large δ^2 values.

Another criterion is the geometric mean of the δ^2 residuals,

$$gM = \exp\left(\frac{1}{n-1} \sum_{i=2}^{n} \ell n\left(\delta_i^2\right) \right). \tag{4.28}$$

Note that

$$\exp\left(\frac{1}{n-1}\sum_{i=2}^{n}\ell n(\delta_i^2)\right) = \left(\prod_{i=2}^{n}\delta_i^2\right)^{\frac{1}{n-1}} \tag{4.29}$$

is the geometric mean of the square residuals, δ_i^2. With this criterion, we have reduced the sensitivity to large values but now the problem becomes oversensitivity to values of δ near zero. Suppose, for instance, that by pure chance a δ residual practically vanishes; this residual determines the value of gM.

To construct more stable (*i.e.*, less affected by random error) criteria, we may use instead of δ_i^2 its kriging variance σ_i^2. Of course, δ_i^2 is a true residual whereas σ_i^2 is a theoretical value; however, according to our model, σ_i^2 is the expected value of δ_i^2. Intuitively, this means that σ_i^2 is representative of δ_i^2—as we have made sure by enforcing Equation (4.26)—but σ_i^2 is more stable than δ_i^2.

If we use this idea with Equation (4.28), we obtain the criterion

$$cR = Q_2 \exp\left(\frac{1}{n-1}\sum_{i=2}^{n}\ell n\left(\sigma_i^2\right)\right), \tag{4.30}$$

which is a good index of the agreement between the model and the data. cR is simply the stabilized geometric mean of the square residuals but is a considerably more reliable indicator than gM; note that we have eliminated the chance of accidentally getting a σ_i^2 value that vanishes because, unless there are redundant measurements, we cannot have zero values of kriging variance.

In summary, an accurate and practical approach is to select the parameters that minimize

$$cR = \exp\left(\frac{1}{n-1}\sum_{i=2}^{n}\ell n\left(\sigma_i^2\right)\right) \tag{4.31}$$

subject to the constraint

$$\frac{1}{n-1}\sum_{i=2}^{n}\frac{\delta_i^2}{\sigma_i^2} = 1. \tag{4.32}$$

A convenient feature of this method is that the estimates do not depend on how the data are ordered. This is because neither (4.31) nor (4.32) depend on the order with which the data are processed. Algorithms are available for the solution of this optimization problem; however, the computational cost may be large if the data set exceeds a couple of hundred measurements, and these algorithms will likely not function well if the problem involves too many or poorly identifiable parameters. In simple applications involving up to two parameters, it is straightforward to optimize the parameters manually.

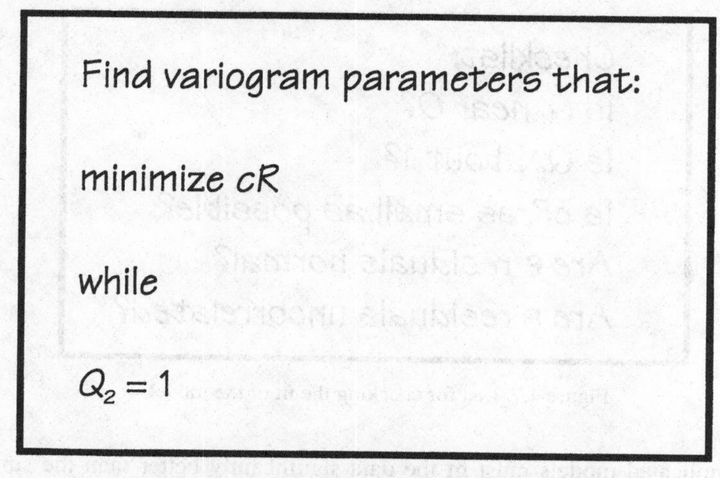

Figure 4.6 Recommended procedure to estimate variogram parameters.

This method, which is summarized in Figure 4.6, is practically the same as the *Restricted Maximum Likelihood* (RML) parameter estimation method of statistical theory ([28, 113]). This method is described in references [79, 81, and 88]. A related method is given in reference [130]. For large samples, this approach has been shown mathematically to yield estimates of the variogram parameters that are unbiased and have the smallest mean square error (given the data). However, the RML method and the asymptotic properties of these parameter estimation methods are beyond the scope of this book. We have used some intuitive arguments to demonstrate that this is a reasonable method if we use an acceptable model. Methods for estimating variograms are also reviewed in references [81 and 149]. A resampling approach is given in reference [16].

4.7 On modeling

It may be worth reiterating that, in practice, a model and its parameters are often selected based on criteria other than their agreement with data: additional information (sometimes called "soft data"), preference of the analyst for certain models, and even available software. However, in this section we will focus on the data set and its use on model selection.

The intrinsic model that has been the subject of Chapters 3 and 4 is the simplest geostatistical model and is also the one most commonly used in practice. There are several reasons that explain its popularity:

1. This first reason is the general principle, known as Occam's razor, that underlies the scientific approach: Use the simplest model that fits the data. More

Checklist:

Is Q_1 near 0?

Is Q_2 about 1?

Is cR as small as possible?

Are ε residuals normal?

Are ε residuals uncorrelated?

Figure 4.7 List for checking the fit of the model.

complicated models must fit the data significantly better than the simple model to assure that they will give better predictions.

2. To apply the intrinsic isotropic model, only the variogram is needed.
3. The intrinsic isotropic model is generally quite a conservative choice, in the sense that to use it one assumes very little about the structure of the function. (The drawback, however, is that a more elaborate model, if justified, could yield estimates with smaller mean square error.)

A general principle is to fit a variogram with as few parameters as possible. Include a nugget-effect variogram if the data indicate clearly the presence of a discontinuity. Include a model with a sill (such as the exponential) only if the data clearly show stationarity; otherwise, use a linear variogram. It is generally quite difficult to estimate from data more than two variogram parameters. In many applications, one can make do with models having only one or two parameters. When comparing a model with three parameters against a model with two parameters, the three-parameter model should be chosen only if it fits the data clearly better than the two-parameter model.

Variogram estimation is an iterative procedure. Several models are fitted and tested. Figure 4.7 gives a list of criteria that must be met to assure that the model fits the data adequately. A variogram that meets the criteria in this list will be sufficient for most practical applications (interpolation, contouring, volume averaging). If more than one model seems to fit the data equally well, select the simplest. In most applications the results from kriging will not be sensitive to the exact shape of the variogram, again assuming that the criteria of Figure 4.7 are met. For example, see reference [14]. A formal approach for incorporating parameter uncertainty into the prediction is described in reference [80]. An approach for incorporating additional information can be found in reference [108].

4.8 Estimation simulator

It takes practice to become good at model selection and calibration. We will practice using an *estimation simulator* (the same idea as a flight simulator), which uses data that were synthetically generated by a mathematical procedure. The idea is that before we try to solve a problem for which the answer is unknown, it is prudent to solve some problems with known answer. Here, we will generate data from a realization of a function with a given variogram and then we will try to infer the variogram from the data. Other more challenging exercises can also be devised, but due to space limitations we will present here only two examples.

A comment is in order on the relevance of such exercises to real-world applications. In practice, there is no such thing as the "true" variogram of the function we want to estimate; the selected variogram represents information about the structure of that function. Thus, the role of estimation simulators is to help us find out whether or how the main structural features can be indentified adequately for practical purposes.

4.8.1 Synthetic realizations

First, we need to make a few comments on generating synthetic realizations. As we discussed in Chapter 3, the intrinsic model with variogram $\gamma(h)$ means that the function of interest, $z(\mathbf{x})$, belongs to an ensemble of many functions (or realizations) and

$$E[z(\mathbf{x}) - z(\mathbf{x}')] = 0 \tag{4.33}$$

$$\frac{1}{2}E[(z(\mathbf{x}) - z(\mathbf{x}'))^2] = \gamma(|\mathbf{x} - \mathbf{x}'|), \tag{4.34}$$

where E denotes averaging over all possible realizations in the ensemble. Now, given $\gamma(h)$, there are computational methods that can generate as many equiprobable realizations as we want. These synthetic realizations are such that they meet statistical tests for model validation. For instance, 95% of them pass the $|Q_1| < \frac{2}{\sqrt{n-1}}$ test. Furthermore, if n realizations are generated, then

$$\frac{1}{n}\sum_{i=1}^{n}[z(\mathbf{x}; i) - z(\mathbf{x}'; i)] \tag{4.35}$$

and

$$\frac{1}{n}\sum_{i=1}^{n}\frac{1}{2}[z(\mathbf{x}; i) - z(\mathbf{x}'; i)]^2 - \gamma(|\mathbf{x} - \mathbf{x}'|) \tag{4.36}$$

both decrease as n increases and tend to zero as n tends to infinity.

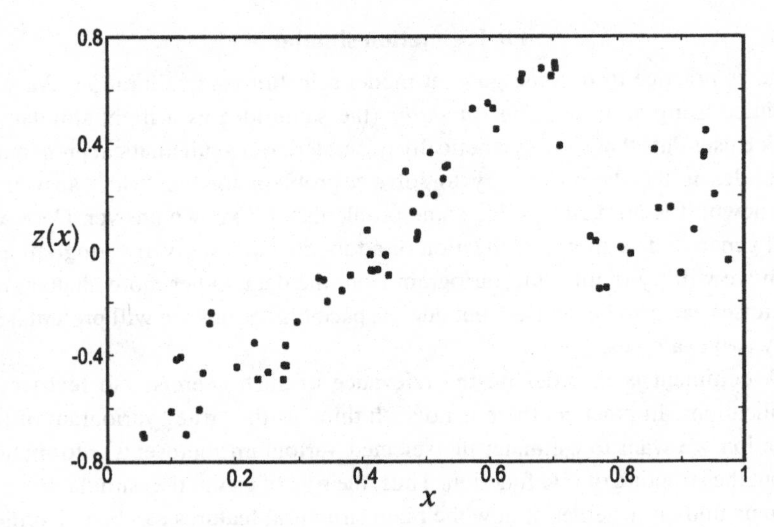

Figure 4.8 Plot of observations versus locations for Example 1.

4.8.2 Example 1

Consider 70 measurements that were synthetically generated from a one-dimensional random process. Our task is to infer the variogram and to test the intrinsic model.

We always start by plotting and calculating basic statistics for the data (exploratory analysis). For one-dimensional data, the task of exploratory analysis is simplified because a plot of the observations versus the location (see Figure 4.8), pretty much conveys a visual impression of the data. In this case, we infer that the data are reasonably continuous but without a well-defined slope. The data indicate clearly that much of the variability is at a scale comparable to the maximum separation distance, so that the function does not appear stationary. These observations are important in selecting the variogram.

Next, we plot the experimental variogram. For a preliminary estimate, we subdivide the separation distances into 10 equal intervals. One might proceed to draw a straight line from the origin through the experimental variogram, as shown in Figure 4.9. That is, our preliminary estimate is that the variogram is linear with slope 0.40.

To test the model and possibly refine the preliminary estimate of the variogram, we compute the residuals. The experimental value of Q_1 is -0.25, and 95% of the values of Q_1 are expected to be within the interval ± 0.25. The experimental value of Q_2 is 2.34 with 95% of the values of Q_2 expected to lie within the interval [0.66, 1.34]. These statistics indicate that something is

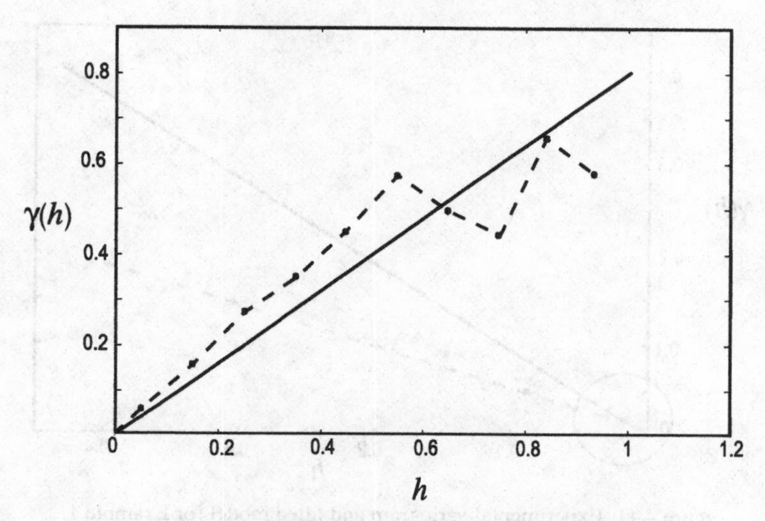

Figure 4.9 Experimental variogram and preliminary fit.

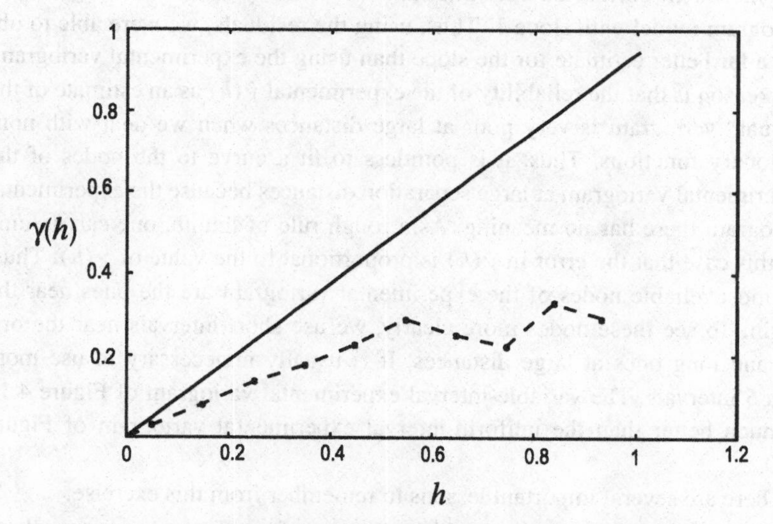

Figure 4.10 Experimental variogram and fit from residuals.

wrong with the model that we fitted. The orthonormal residuals clearly do not have variance 1, and they likely do not have mean 0.

The best fit obtained using the method of Figure 4.6 yields a slope of 0.94. Plotting this fitted model with the experimental variogram in Figure 4.10, we see that the two do not seem to be in great agreement. (Actually, as we will see, the fitted model and the variogram are in agreement where it matters.)

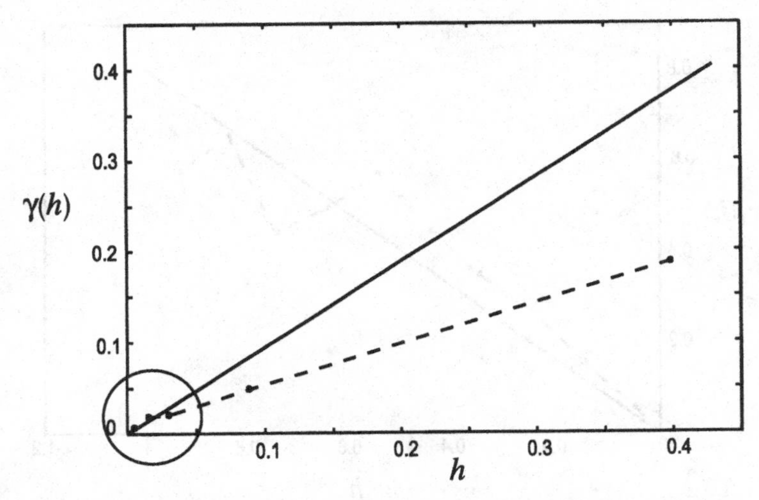

Figure 4.11 Experimental variogram and fitted model for Example 1.

Which is the correct answer? The data were actually generated using a linear variogram model with slope 1. Thus, using the residuals, we were able to obtain a far better estimate for the slope than using the experimental variogram. The reason is that the reliability of the experimental $\hat{\gamma}(h)$ as an estimate of the "actual" variogram is very poor at large distances when we deal with nonstationary functions. Thus, it is pointless to fit a curve to the nodes of the experimental variogram at large separation distances because the experimental variogram there has no meaning. As a rough rule of thumb, one can assume for this case that the error in $\hat{\gamma}(h)$ is proportional to the value of $\gamma(h)$. Thus, the most reliable nodes of the experimental variogram are the ones near the origin. To see these nodes more clearly, we use short intervals near the origin and long ones at large distances. It is usually unnecessary to use more than 5 intervals. The variable-interval experimental variogram of Figure 4.11 is much better than the uniform-interval experimental variogram of Figure 4.10.

There are several important lessons to remember from this exercise:

1. If the function is continuous (which we can see from the exploratory analysis of the data as well as from the behavior of the experimental variogram near the origin), then fit a variogram that reproduces the experimental variogram quite closely near the origin. Compute an experimental variogram using short intervals near the origin, as shown in Figure 4.11.
2. If the function is nonstationary (which we can often infer from the exploratory analysis of the data), pay no attention to the values of the experimental variogram at large distances. Typically, for a function that is both

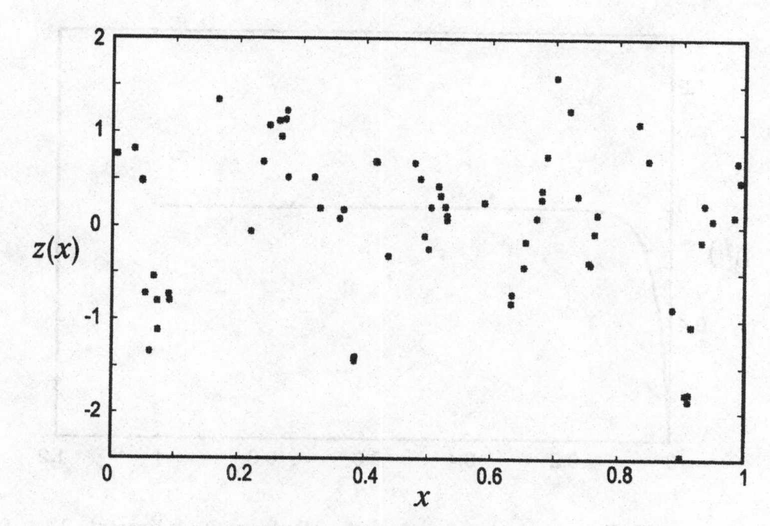

Figure 4.12 Data for Example 2.

nonstationary and continuous, disregard the variogram points at distances larger than half the maximum separation distance.

3. It can be shown[1] that the relative standard error of estimation of the slope of the linear variogram is $\sqrt{2/(n-1)}$; thus, the larger the sample, the smaller the relative error.

4.8.3 Example 2

Again we deal with 70 measurements of a one-dimensional function. Let us plot the data in Figure 4.12. From this plot, we see that the function from which we have data may be stationary (*i.e.*, large-scale components are not very prominent) and perhaps the function is continuous.

We next plot the experimental variogram using two different discretizations, shown in Figures 4.13 and 4.14. The second discretization shows that the function is indeed practically continuous. We choose an exponential variogram,

$$\gamma(h) = v\left(1 - \exp\left(-\frac{h}{l}\right)\right), \tag{4.37}$$

with preliminary estimates $v = 0.8$ and $l = 0.03$. Next, we try to refine the estimate by adjusting the value of l to optimize the value of cR.

[1] This result and some others mentioned later in this section are based on the method of restricted maximum likelihood ([79, 88, 81]). A number of implicit assumptions are made, such as normality and large sample size.

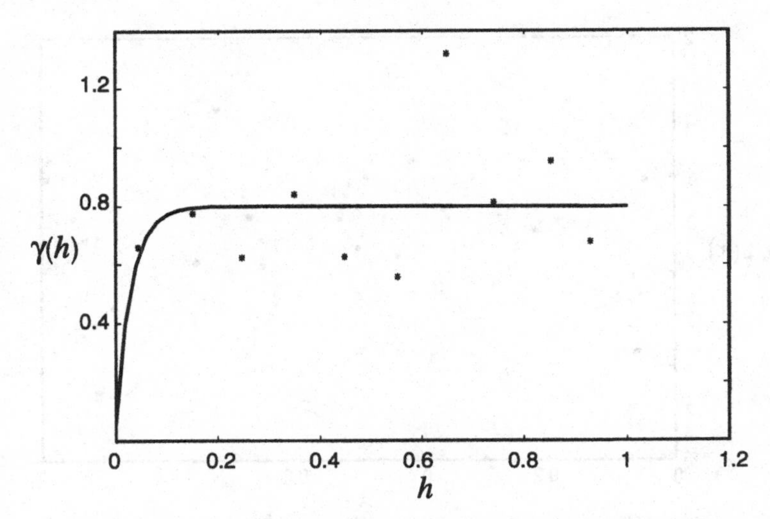

Figure 4.13 Experimental variogram and preliminary fit for Example 2.

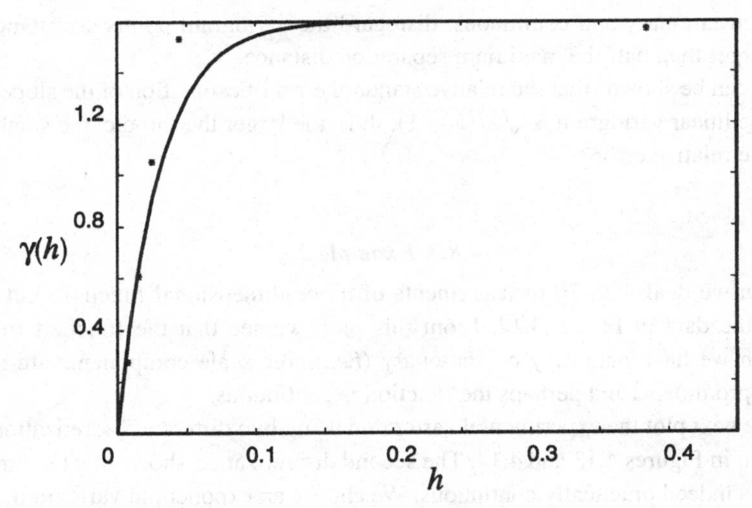

Figure 4.14 Experimental variogram and preliminary fit for Example 2 using another discretization.

We see that the fitting criterion is quite insensitive to the value of l (see Table 4.3). That is, good fits can be obtained for a range of values of l. The estimate that gives the smallest cR is $l = 0.03$ and $v = 0.64$ (to make $Q_2 = 1$). This model is shown in Figure 4.15. For these parameters, Q_1 is 0.12, which means that the average of the residuals is insignificantly different from zero; the residuals pass the normality test (see Figure 4.16); and the residuals seem

Table 4.3.
Optimizing the fit

l	cR
0.020	0.2111
0.025	0.2064
0.030	0.2053
0.035	0.2059
0.040	0.2071

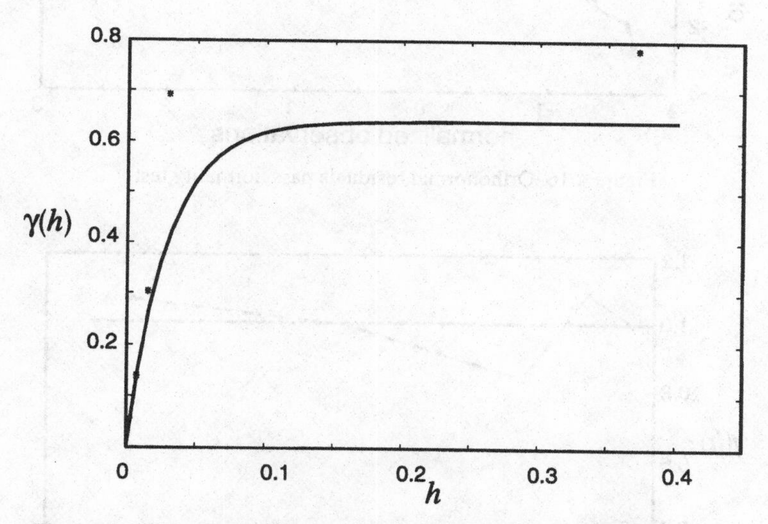

Figure 4.15 Experimental variogram and best fit estimate of variogram.

uncorrelated, as evidenced by their experimental variogram, which is practically a nugget effect with variance 1 (see Figure 4.17).

Thus, we conclude that the variogram is exponential with sill 0.64 and length parameter 0.03. How did we do? (The envelope, please!) The data were generated from a stationary function with exponential variogram with sill 1 and length parameter 0.05. Our estimation failed to come very close to the actual values of the sill and the length parameter, but notice that it did a good job estimating the slope of the variogram at the origin. (Compare "true" value $\frac{1}{0.05} = 20$ with estimate $\frac{0.64}{0.03} = 21$). Again, we see that, because there was no nugget effect, we could reproduce the behavior at the origin. Finding the sill and the length parameter, however, is always much tougher. Fortunately, interpolation and averaging depend mostly on the behavior of the variogram near the origin.

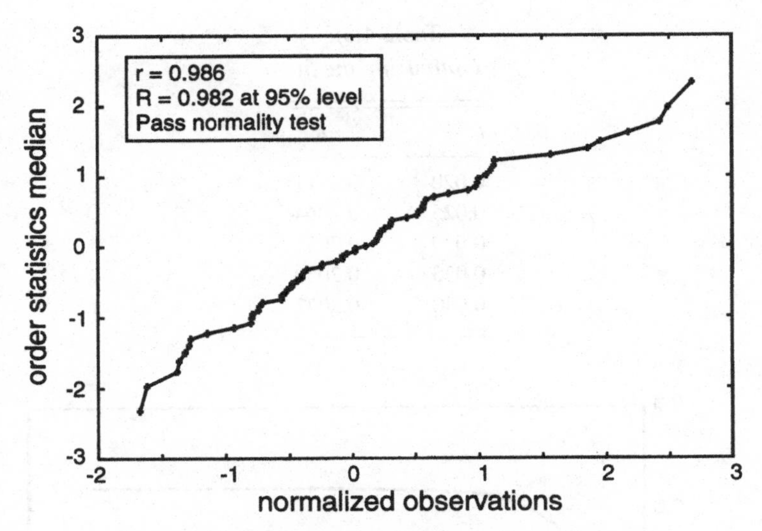

Figure 4.16 Orthonormal residuals pass normality test.

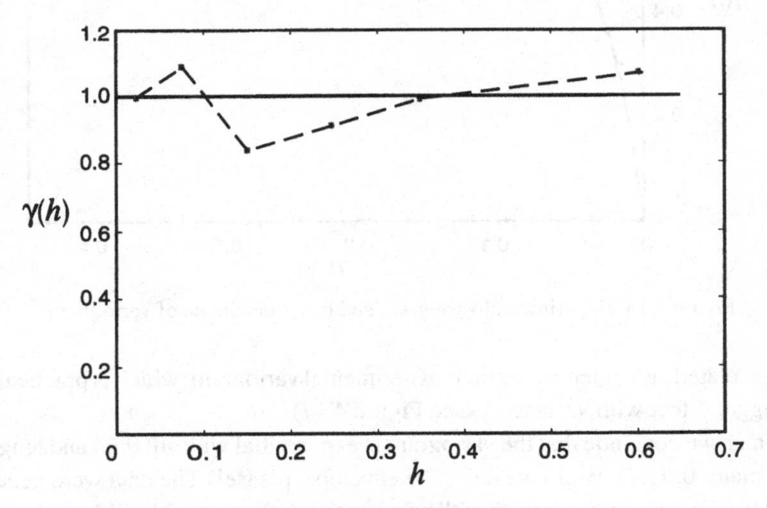

Figure 4.17 Experimental variogram of residuals in nugget effect.

Some lessons can be learned from this exercise:

1. If the function is continuous, then fit a variogram that reproduces the experimental variogram quite closely near the origin.
2. Even if a function is stationary, the errors associated with estimation of the sill and the length parameter may be large and are positively correlated (*i.e.*, these

parameters are usually either both overestimated or both underestimated). The relative error in the sill turns out to be of the order of $\sqrt{2(l/L)}$, where l is the length parameter and L is the size of the sampling domain, provided that we have sufficient measurements. In this example, we expect that the relative error will be of the order of 30%.

3. It can be shown that for a continuous stationary function, the slope of the variogram at the origin can be inferred with relative standard error of the order of $\sqrt{2/(n-1)}$ or less. Thus, for large samples, we can expect to obtain accurate estimates of the slope of the variogram, which is good news if we intend to solve problems of interpolation or averaging.

4.9 Key points of Chapter 4

We examined the problem of fitting a variogram to the data and testing the efficacy of the model. In practically all applications, data are insufficient for estimating accurately the whole variogram. However, a reasonable model can be obtained following an iterative procedure. Starting with the experimental variogram, a model is postulated, which is then improved using residuals. The recommended procedure for variogram fitting relies more on the residuals than on the experimental variogram. "Fitting a (model) variogram" should not be confused with fitting to the experimental variogram; instead, fitting a variogram should be to the data in the sense that residuals are small and have the properties anticipated by the model. Figure 4.7 summarizes a practical way to evaluate the model fit.

5
Anisotropy

The structure of a variable may depend on the direction. This chapter provides an overview of selected techniques for modeling such behavior.

5.1 Examples of anisotropy

The variograms studied in Chapters 3 and 4 are examples of *isotropic* models, in which the correlation structure (in particular, the variogram) does not differ with the orientation. The variogram depends only on the separation distance h and not on the orientation of the linear segment connecting the two points; that is, the average square difference between two samples at distance h is the same whether this distance is in the horizontal or in the vertical direction.

There are many cases, however, where the structure of a certain variable does depend on the direction (*anisotropy*). The best examples of anisotropic structures can be found in stratified formations. Consider, for example, an alluvial unit formed by the deposition of layers of gravel, sand, and silt. The layers and lenses of the various materials are oriented in the horizontal direction; as a result, the mean square difference of hydraulic conductivity measured at two points at a short distance h in the horizontal direction is smaller than the mean square difference at the same distance in the vertical direction. The same holds true for other hydrogeologic parameters, such as storativity.

Figure 5.1 is a vertical cross section showing a sketch of the cross section of an alluvial formation. Assume that we compute the experimental variogram γ_1 as a function only of the separation distance h_1 in the vertical direction and then we compute the experimental variogram γ_2 as a function of the horizontal distance h_2. (We use the same method as for the experimental variogram except that, instead of the overall distance, we use the distance in the horizontal or vertical direction.) The two variograms, shown in Figure 5.2, have the same sill, but γ_2 reaches the sill at a shorter separation distance than does γ_1. That is,

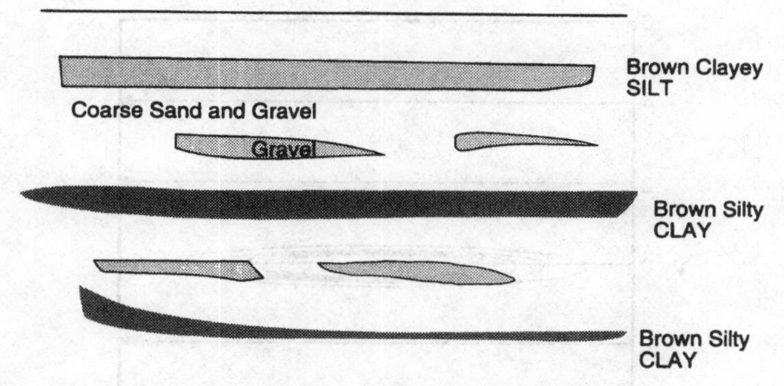

Figure 5.1 Vertical cross section of alluvial formation.

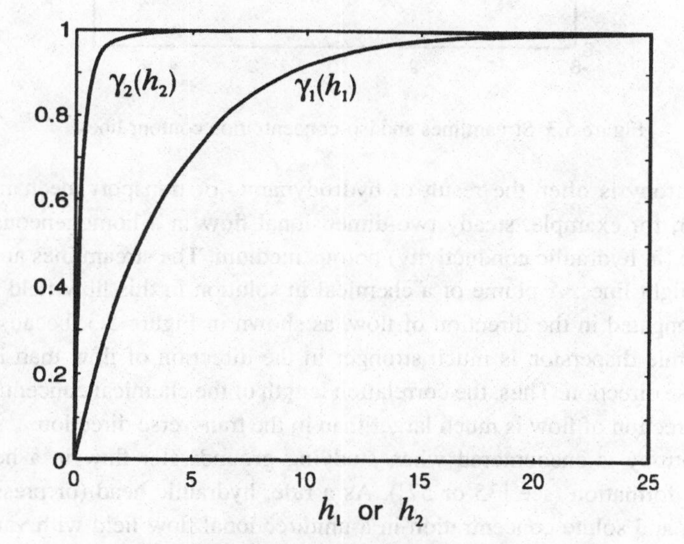

Figure 5.2 Variograms in horizontal and vertical directions.

the correlation distance in the vertical direction is shorter than in the horizontal direction.

Consider the practical implications of anisotropy in, for example, interpolation. From measurements of log-conductivity (logarithm of hydraulic conductivity) in boreholes, we want to estimate the log-conductivity at another point. A measurement at 10 m horizontal distance is probably as informative (*i.e.*, should be given the same weight) as a measurement at 1 m vertical distance because there is much more correlation in the horizontal direction than there is in the vertical direction.

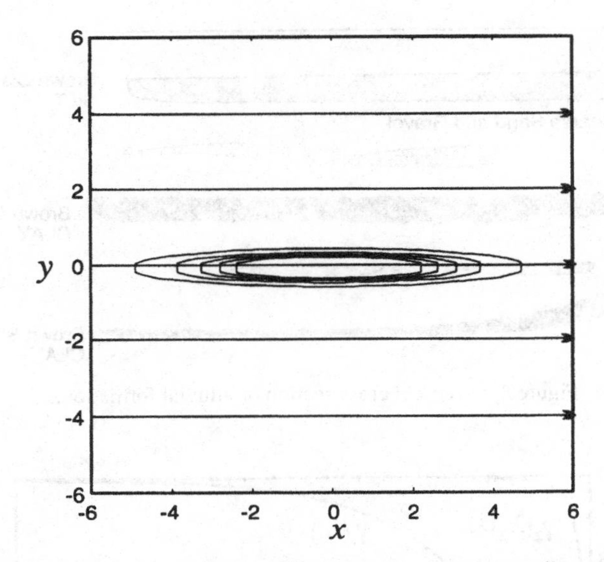

Figure 5.3 Streamlines and iso-concentration contour lines.

Anisotropy is often the result of hydrodynamic or transport mechanisms. Consider, for example, steady two-dimensional flow in a homogeneous and isotropic (in hydraulic conductivity) porous medium. The streamlines are parallel straight lines. A plume of a chemical in solution in this flow field tends to be elongated in the direction of flow, as shown in Figure 5.3, because hydrodynamic dispersion is much stronger in the direction of flow than in the transverse direction. Thus, the correlation length of the chemical concentration in the direction of flow is much larger than in the transverse direction.

Anisotropy is encountered when studying groundwater flow in a heterogeneous formation (see [35 or 57]). As a rule, hydraulic head (or pressure), velocity, and solute concentration in a unidirectional flow field with variable conductivity are anisotropic; that is, the variogram in the direction of flow is different from the variogram in the direction perpendicular to it. The "principal axes" of the anisotropy are determined by the direction of flow.

Finally, examples of anisotropic structures are encountered in hydrometeorology. The structure of precipitation is often anisotropic as the result of wind prevailing in a certain direction or orographic effects.

5.2 Overall approach

In many of the cases mentioned above, such as conductivity in a layered medium or concentration in a flow field, we can probably develop a model that represents available information better and that has the potential to give more accurate predictions by accounting for anisotropy. A drawback of anisotropic models is that they require more parameters than the corresponding isotropic models. Consequently, anisotropic models must be used only when sufficient data or prior information allow us to differentiate with respect to direction.

The overall approach used with anisotropic intrinsic models is the same as the approach for isotropic intrinsic models that we saw in Chapters 3 and 4. The basic difference is that in the anisotropic case the variogram depends on the separation vector $\mathbf{h} = [h_1, h_2, h_3]$, instead of the scalar separation distance $h = \sqrt{h_1^2 + h_2^2 + h_3^2}$, known as the Euclidean distance, that is used in the isotropic case. In other words,

> in the anisotropic case, the separation between two points is characterized not only by distance but also by orientation.

Consequently, the choice of a system of spatial coordinates is more important than it was in the isotropic case. An example of an anisotropic variogram is

$$\gamma(h_1, h_2, h_3) = \sigma^2 \left[1 - \exp\left(-\sqrt{\left(\frac{h_1}{20}\right)^2 + \left(\frac{h_2}{20}\right)^2 + \left(\frac{h_3}{2}\right)^2} \right) \right], \quad (5.1)$$

where h_1 and h_2 are distances along the horizontal axes of a Cartesian system and h_3 is the distance along the vertical axis, measured in meters. This variogram conveys the information that the correlation length in a horizontal direction is ten times longer than the correlation length in the vertical direction.

Using linear estimation (kriging) in an anisotropic case is as easy as in an isotropic case. Exactly the same formulae apply, with the exception that we use $\gamma(h_1, h_2, h_3)$ or $\gamma(\mathbf{h})$ instead of $\gamma(h)$. The challenge lies in structural analysis, because anisotropic variograms have more parameters and are harder to estimate. We follow the usual approach in which a variogram is tentatively selected and then tested using the methods that we saw in Chapter 4. It is important to keep in mind that, in practice, one should adopt a complicated empirical model only if it fits the data significantly better than a simpler model: In other words, always use the simplest empirical model consistent with the data.

5.3 Directional variogram

The *directional experimental variogram* is a tool useful in exploring the degree of anisotropy of the data. To plot it, consider the raw variogram. For two dimensions, in addition to the separation distance, compute the direction angle ϕ_{ij} (where angles differing by 180° are considered the same, *e.g.*, 110° is the same as −70°) for any pair of measurements. The data pairs are grouped with respect to the orientation. Then a separate directional experimental variogram is plotted for each group of direction angles. That is, plot one experimental variogram for those in the interval $-90°$ to $0°$ and another experimental variogram for those in the interval $0°$ to $90°$. Commonly, in two-dimensions, two orthogonal directions (those whose orientations differ by $90°$) are selected.

Significant differences between directional variograms may be indicative of anisotropy, and the orthogonal directions where the contrast is the sharpest may be taken as the principal directions of anisotropy. However, the same differences may be indicative of a drift. Thus, the directional variogram is an exploratory tool that can be used to suggest models worthy of further analysis. However, the next step is to propose a model and to calibrate and validate it using orthonormal residuals.

5.4 Geoanisotropy

5.4.1 General

After data exploration, we tentatively select a model. The most useful type of empirical model is one in which the variable is isotropic in an appropriately chosen Cartesian coordinate system. Consider, for example, the anisotropic variogram model of Equation (5.1). We can transform from the original (x_1, x_2, x_3) coordinate system to the new system (x_1^*, x_2^*, x_3^*):

$$x_1^* = x_1, \quad x_2^* = x_2, \quad x_3^* = 10x_3 \tag{5.2}$$

$$\gamma(h_1, h_2, h_3) = \sigma^2\left[1 - \exp\left(-\sqrt{\left(\frac{h_1^*}{20}\right)^2 + \left(\frac{h_2^*}{20}\right)^2 + \left(\frac{h_3^*}{20}\right)^2}\right)\right]$$

$$= \sigma^2\left[1 - \exp\left(-\frac{h^*}{20}\right)\right], \tag{5.3}$$

where $h^* = \sqrt{(h_1^*)^2 + (h_2^*)^2 + (h_3^*)^2}$.

Thus, the simple "stretching" of the vertical axis allows us to use a familiar isotropic model. Whereas in this example we used an exponential variogram, any of the variograms that we saw in Chapter 3 can be used. Also, it should be clear that the same approach applies if we work with covariance functions instead of variograms. In the remainder of this section we will discuss this useful class of models, making references to the variogram of an intrinsic function.

The basic premise is that there is a special rectangular Cartesian coordinate system, which can be obtained from the original system through axis rotation and stretching, in which the variogram depends only on the Euclidean distance. The axes of this special system are known as *principal*. The term *geoanisotropy*

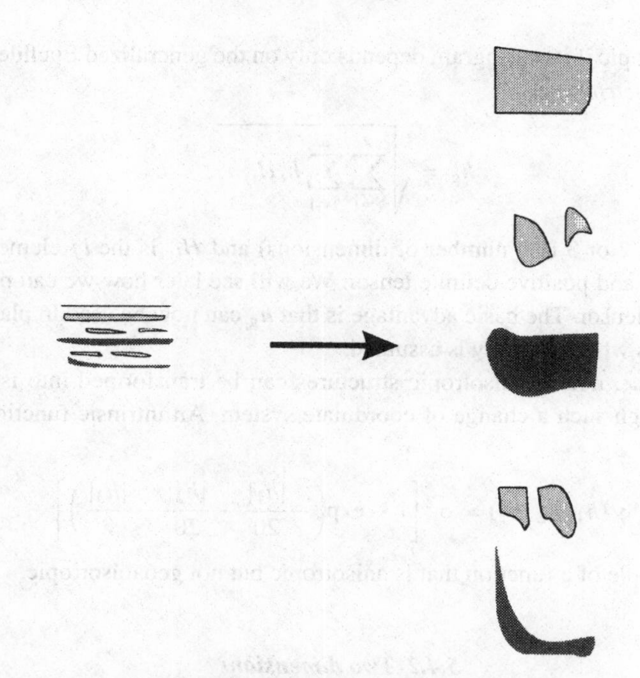

Figure 5.4 Vertical dimension stretched by a factor 12.

has been coined to describe this specific model of anisotropy. (The term *geometric anisotropy*, used in other books, refers to the more general model where the anisotropy can be reduced to isotropy by a linear transformation of coordinates. Geoanisotropy limits the choice of linear transformation to a combination of axis rotations and stretchings.)

It is easier to visualize the concept of geoanisotropy in the two-dimensional case of a vertical cross section in a stratified formation (see Figures 5.1 and 5.4). The original coordinate system is (x_1, x_2). First, we rotate the axes to align them with the directions parallel and perpendicular to the stratification, which are the directions of maximum and minimum correlation length, *i.e.*, the principal axes. Then, we stretch the domain in the direction of minimum correlation length (perpendicular to the stratification) by a coefficient larger than 1. In this new coordinate system, the variogram depends only on the separation distance.

In three dimensions, the same general approach applies. The transformation generally involves three rotation angles, needed to align the coordinate system with the principal directions, and two stretching parameters, needed to equalize the correlation length in all three dimensions.

The same idea can be expressed in a mathematically succinct way using the concept of a *generalized distance*. We say that an intrinsic function is geoanisotropic if its variogram depends only on the generalized Euclidean distance (or *metric*):

$$h_g = \sqrt{\sum_{i=1}^{d} \sum_{j=1}^{d} h_i H_{ij} h_j}, \tag{5.4}$$

where d is 2 or 3 (the number of dimensions) and H_{ij} is the ij element of a symmetric and positive definite tensor. We will see later how we can parameterize this tensor. The basic advantage is that h_g can now be used in place of h in methods where isotropy is assumed.

Of course, not all anisotropic structures can be transformed into isotropic ones through such a change of coordinate system. An intrinsic function with variogram

$$\gamma(h_1, h_2, h_3) = \sigma^2 \left[1 - \exp\left(-\frac{|h_1|}{20} - \frac{|h_2|}{20} - \frac{|h_3|}{2} \right) \right] \tag{5.5}$$

is an example of a function that is anisotropic but not geoanisotropic.

5.4.2 Two dimensions

Let us first study the transformation to the principal-axes system:

$$\begin{aligned} x_1^* &= \cos(\varphi)x_1 + \sin(\varphi)x_2 \\ x_2^* &= -\alpha \sin(\varphi)x_1 + \alpha \cos(\varphi)x_2, \end{aligned} \tag{5.6}$$

where, as Figure 5.5 indicates, φ is the rotation angle needed to bring positive x_1 to overlap with positive x_1^* and α is the stretching applied on the second axis

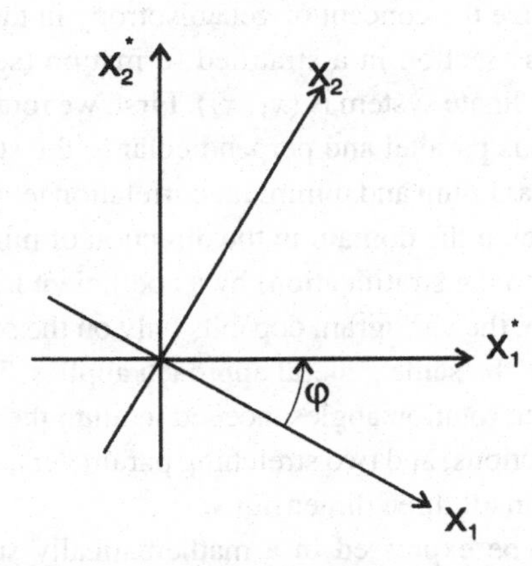

Figure 5.5 Rotation of coordinate system (new system is denoted by *).

after the rotation. The angle φ is positive in the counterclockwise direction. Note that both systems are assumed positively oriented, *i.e.*, if we rotate the first axis by 90° in the counterclockwise direction we make it coincide with the second. In vector notation,

$$\begin{bmatrix} x_1^* \\ x_2^* \end{bmatrix} = \begin{bmatrix} 1 & 0 \\ 0 & \alpha \end{bmatrix} \begin{bmatrix} \cos(\varphi) & \sin(\varphi) \\ -\sin(\varphi) & \cos(\varphi) \end{bmatrix} \begin{bmatrix} x_1 \\ x_2 \end{bmatrix}$$

or

$$\mathbf{x}^* = \mathbf{M}\,\mathbf{T}\,\mathbf{x},$$

where $\mathbf{T}$ is the matrix of rotation and $\mathbf{M}$ is the diagonal matrix of stretching.

Thus, this transformation involves two user-chosen parameters: the rotation angle φ and the stretching parameter α. Tensor $\mathbf{H}$ is parameterized as follows:

$$\mathbf{H} = \begin{bmatrix} \cos^2\varphi + \alpha^2 \sin^2\varphi & (1-\alpha^2)\sin\varphi\cos\varphi \\ (1-\alpha^2)\sin\varphi\cos\varphi & \sin^2\varphi + \alpha^2\cos^2\varphi \end{bmatrix}, \tag{5.7}$$

where φ is an angle between $-90°$ and $90°$ and α is a positive parameter.

Let us consider two special cases:

Isotropic In this case $\alpha = 1$, so that, from the previous equation,

$$\mathbf{H} = \begin{bmatrix} 1 & 0 \\ 0 & 1 \end{bmatrix}, \tag{5.8}$$

and the generalized Euclidean distance reduces to the usual separation distance. The value of the rotation angle φ does not matter because it has no effect on the distance. That is, in the isotropic case, the orientation of the coordinate system is unimportant.

Anisotropic but aligned with principal axes In this case $\varphi = 0$, so that

$$\mathbf{H} = \begin{bmatrix} 1 & 0 \\ 0 & \alpha^2 \end{bmatrix}, \tag{5.9}$$

which is equivalent to simple stretching of one of the axes.

5.4.3 Three dimensions

Not surprisingly, the equations for rotation in three dimensions are considerably more complex. We will summarize in vector notation key results that may be useful to some of the readers. We assume that all systems are positively oriented: By convention, a system $x_1x_2x_2$ is positively oriented if when viewed from an arbitrary point on the positive x_3 semiaxis, the positive x_1 axis would have to be rotated counterclockwise by 90° to coincide with the positive x_2 axis.

We will consider the rotation from an original coordinate system $x_1x_2x_2$ to a new coordinate system $x_1^*x_2^*x_2^*$. The transformation in vector notation is

$$\mathbf{x}^* = \mathbf{T}\,\mathbf{x}, \tag{5.10}$$

where T_{ij} is the cosine of the angle formed between the positive semiaxes x_i^* and x_j. Formula (5.10) constitutes all that is needed.

To verify that **T** is indeed such a matrix, the matrix logarithm of **T** must be a skew symmetric matrix:

$$logm(\mathbf{T}) = \boldsymbol{\varphi} = \begin{bmatrix} 0 & \varphi_3 & -\varphi_2 \\ -\varphi_3 & 0 & \varphi_1 \\ \varphi_2 & -\varphi_1 & 0 \end{bmatrix}. \tag{5.11}$$

That is, because of the constraints that must be satisfied by the nine cosines, the transformation matrix is defined fully through three variables: the rotation angles φ_1, φ_2, and φ_3, which are positive if directed in the counterclockwise direction. Here, φ_1 is the rotation about axis x_1, etc. Sometimes, it is more convenient to define these three angles and to find **T** from the matrix exponential:

$$\mathbf{T} = expm(\boldsymbol{\varphi}). \tag{5.12}$$

See reference [131] regarding the matrix exponential, *expm*, and the matrix logarithm, *logm*. Most computer packages have efficient algorithms to compute the matrix exponential and the matrix logarithm. (For example, in MATLAB, these are functions `expm` and `logm`.)

5.5 Practical considerations

Unless the form of the variogram is known, such as from stochastic theory, as in references [35 or 57], we may limit our attention to geoanisotropic models. This means that there exists a special coordinate system where the anisotropic case simplifies to the familiar isotropic case, *i.e.*, only a scalar distance matters. Actually, we do not even need to carry out the transformation; all we have to do is replace in the expression for the variogram (or covariance function) the Euclidean distance h by the metric h_g given by Equation (5.4). Thus, the modeling of anisotropy boils down to determination of **H**.

The simplest case is when the axes have been chosen so that they are already aligned with the principal directions. This proves to be relatively straightforward because in most applications the directions of anisotropy are obvious (direction of overall stratification, direction of mean flow, etc.). In this case, the only parameters are the stretching coefficients. This means that the modeler needs to choose only one stretching coefficient in the two-dimensional case and two coefficients in the three-dimensional case. In three dimensions, it is common that the structure is isotropic on a plane; for example, in the case of stratifications, the correlation structure is isotropic in the two directions parallel to the stratification but differs from the structure perpendicular to it, as in Equation (5.1). Consequently, only one stretching parameter needs to be computed.

If the axes of the system that gives the coordinates of the points do not coincide with the principal directions of anisotropy, but these directions are known, then one needs to rotate the axes. This is simple in two dimensions because all one needs to specify is the rotation angle. In three dimensions, however, visualization poses some difficulty . Possible methods to specify the rotation include: (*a*) specifying the nine directional cosines (the cosines of each principal axis with respect to each presently used axis), a straightforward but tedious approach, and (*b*) computing the nine angles from three of them.

Adjustment of parameters can be made based on the minimization of the *cR* fitting criterion. However, blindly trying to adjust the parameters of the general geoanisotropic model is not warranted in most applications, especially in three-dimensions where it is hard to justify five adjustable parameters in addition to the parameters of the isotropic model. It is best to fix as many of the parameters as possible based on process understanding and to leave only a few parameters for fitting to the data.

Exercise 5.1 *Assume that you have a data set of a quantity that varies in two-dimensional space and that we consider two models:*

(a) an isotropic model with linear plus nugget-effect variogram.
(b) an anisotropic model with linear plus nugget-effect variogram.

Justify that the minimum cR (obtained by optimizing the parameters so that they minimize cR) for model (b) will always be smaller than or equal to the value of the minimum cR for model (a). Then, suggest a procedure that can be used to determine whether the difference in cR between the two models is large enough to justify use of the more complicated model (b). Just recommend the procedure without being concerned about analytical or computational difficulties in the implementation.

5.6 Key points of Chapter 5

When the correlation structure of a hydrogeologic property varies with the direction, as in the case of layered formations, it is possible to utilize more information and thus improve estimation accuracy by taking this anisotropy into account. In the absence of information pointing to another model, the geoanisotropic model may be used: The correlation may be transformed to isotropic through rotation and stretching of the axes.

6
Variable mean

In this chapter, we justify the use of models with variable mean and show how the methods we saw in the previous chapters for constant-mean cases can be extended. We also discuss generalized covariance functions, which take the place of variograms.

6.1 Limitations of constant mean

In Chapters 3–5, we dealt with the intrinsic model. This model represents the variable of interest, $z(\mathbf{x})$, as wavering about a constant value, and the variogram provides information about the scale and intensity of fluctuations. The fluctuations are characterized statistically as having zero mean and a certain correlation structure. In more mathematical terms, if $z(\mathbf{x})$ is intrinsic, then

$$z(\mathbf{x}) = m + \epsilon(\mathbf{x}), \tag{6.1}$$

where m is a constant and $\epsilon(\mathbf{x})$ is a stochastic process with zero mean,

$$E[\epsilon(\mathbf{x})] = 0, \tag{6.2}$$

and variogram

$$\frac{1}{2}E[(\epsilon(\mathbf{x}) - \epsilon(\mathbf{x}'))^2] = \gamma(\mathbf{x} - \mathbf{x}'). \tag{6.3}$$

One way to interpret Equation (6.1) is to think of $z(\mathbf{x})$ as consisting of a *deterministic* part, $z_d(\mathbf{x})$, and a *stochastic* part, $z_s(\mathbf{x})$. These two parts differ because the deterministic part involves exact determination and the stochastic part involves approximate correlation: Specification of the parameter m completely determines the value of $z_d(\mathbf{x})$, whereas specification of $\gamma(|\mathbf{x} - \mathbf{x}'|)$ gives an approximate value of $z_s(\mathbf{x})$ based on its correlation with the measurements. In other words, the deterministic part is "definite" whereas the stochastic part is generally "fuzzy." In this sense, the assumptions about the deterministic part

Table 6.1. *Data for illustative example*

Observation	Location
0.20	0.20
0.30	0.30
0.40	0.40
0.50	0.50
0.60	0.60
0.70	0.70
0.80	0.80
1.00	1.00

are stronger than the assumptions about the stochastic part; that is, assumptions about the deterministic part have more impact on the predictions and are more important to justify than assumptions about the stochastic part. If an analyst does not know or is unwilling to assume much about $z(\mathbf{x})$, it is natural to rely more on the stochastic part than on the deterministic part to describe the spatial structure of the variable of interest.

The deterministic part in the intrinsic model is the simplest possible: a constant. This model relies mainly on the correlation structure as expressed by the variogram. In a way, it is a conservative and cautious model; such characteristics explain its popularity in fields where processes are so complex that one cannot confidently make strong assumptions. The intrinsic model often is an adequate representation of the data and suffices for the solution of interpolation and averaging problems. However, there certainly exist cases, especially in hydrogeology, where one is justified in shifting onto the deterministic part some of the burden of representing the spatial variability of $z(\mathbf{x})$.

Let us start with a simple example that will shed light on the strengths and limitations of the intrinsic model. The hypothetical data of Table 6.1 show a clear linear trend. Let us compare two models.

Model 1 In this model the deterministic equation,

$$z = x, \tag{6.4}$$

fits the data perfectly and the predictions are presumed to be error-free.

Model 2 Our second model is intrinsic with power variogram

$$\gamma(h) = \frac{1}{2}h^{1.99}. \tag{6.5}$$

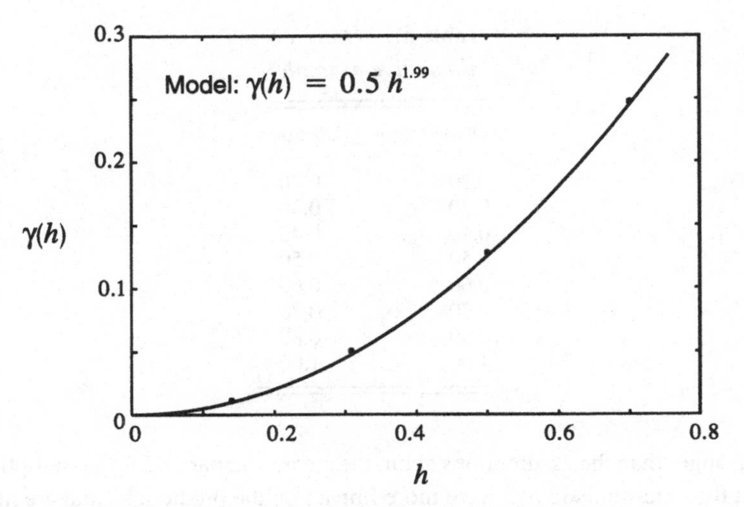

Figure 6.1 Experimental and model variogram for intrinsic case.

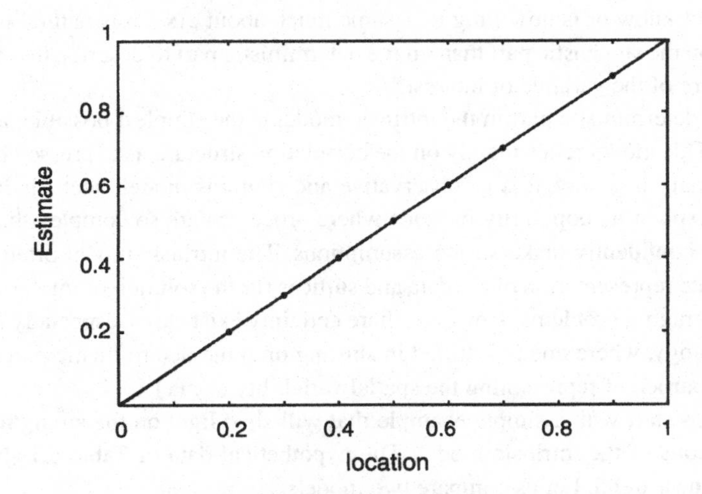

Figure 6.2 Comparison of best estimates.

See Figure 6.1 for a comparison between the experimental and model variogram. (Of course, by now we know that despite the appearance of Figure 6.1, this model does *not* necessarily represent the data well. We will use this model only for illustration purposes.)

Figure 6.2 compares the best estimates using the two models; Figure 6.3 compares the mean square error. The best estimates given by the two models are the same, with some very small differences away from measurements. However, the mean square errors evaluated by the two models are quite different.

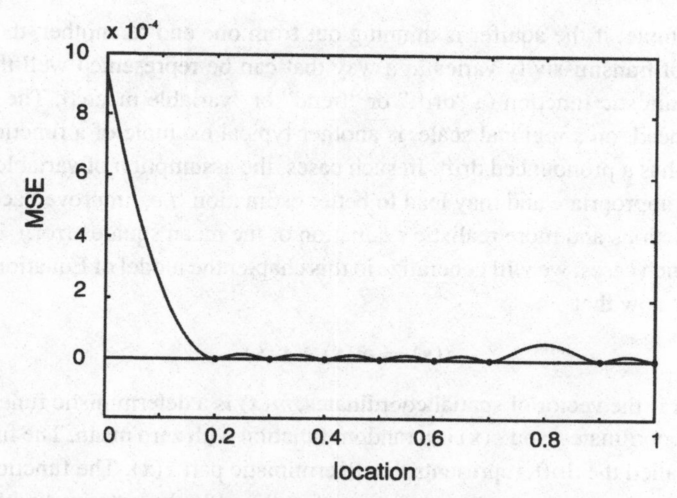

Figure 6.3 Comparison of mean square error of estimation.

In applications, we are often interested in comparing two plausible models that differ in the emphasis that they place on the deterministic part. The two models described in this section represent two opposite extremes: Model 1 relies only on the deterministic part, whereas Model 2 relies as much as possible on the stochastic part. We may draw conclusions from this simple example that are generally applicable:

1. The differences in the best estimates are most pronounced away from measurements. Near measurements, the two methods usually give similar best estimates because of the effect of conditioning on the data.
2. Model 2, which relies mainly on the stochastic part, is more conservative in the evaluation of the accuracy of predictions, *i.e.*, it tends to calculate a larger mean square error than the more self-confident Model 1.

Which model is best? Obviously, the answer depends on our confidence in Equation (6.4). If we have reasons to believe that the good fit that Equation (6.4) has with the data is not coincidental, then the intrinsic model of Equation (6.5) overestimates the mean square error. In Section 6.14, we will see an approach for evaluating whether the data support the use of a variable-mean model.

6.2 The linear model

In some cases, such as the representation of the hydrogeologic parameters of a regional aquifer, there may exist a large-scale component of spatial variability that can be represented with reasonable assurance as a deterministic function.

For example, if the aquifer is thinning out from one end to another, the coefficient of transmissivity varies in a way that can be represented well through a deterministic function (a "drift" or "trend" or "variable mean"). The piezometric head, on a regional scale, is another typical example of a function that usually has a pronounced drift. In such cases, the assumption of variable mean is more appropriate and may lead to better estimation (*i.e.*, improved accuracy of predictions and more realistic evaluation of the mean square error.)

For such cases, we will generalize in this chapter the model of Equation (6.1). Assume now that

$$z(\mathbf{x}) = m(\mathbf{x}) + \epsilon(\mathbf{x}), \tag{6.6}$$

where $\mathbf{x}$ is the vector of spatial coordinates, $m(\mathbf{x})$ is a deterministic function of spatial coordinates, and $\epsilon(\mathbf{x})$ is a random function with zero mean. The function $m(\mathbf{x})$, called the drift, represents the deterministic part $z(\mathbf{x})$. The function $\epsilon(\mathbf{x})$ represents the stochastic ("fuzzy") part of $z(\mathbf{x})$ and is characterized with some sort of correlation function.

For the sake of convenience, the drift is represented by the summation of known functions with unknown coefficients,

$$m(\mathbf{x}) = \sum_{k=1}^{p} f_k(\mathbf{x})\beta_k, \tag{6.7}$$

where $f_1(\mathbf{x}), \ldots, f_p(\mathbf{x})$ are known functions of the spatial coordinates $\mathbf{x}$ and are called trial or base functions. The coefficients $\beta_1, \ldots, \beta_p$ are deterministic but unknown and are referred to as *drift coefficients*. Examples of trial functions are polynomials (see example in next section), trigonometric functions, and families of orthogonal functions (such as sines and cosines). Because the mean function is linear in the drift coefficients, Equations (6.6) and (6.7) constitute what is known as the *linear model* (not to be confused with the linear variogram or the linear trend). The linear model plays a prominent role not only in geostatistics but also in every other field of applied statistics.

Example 6.1 *For z defined on one dimension,*

$$z(x) = \beta_1 + x\,\beta_2 + \epsilon(x) \tag{6.8}$$

means that $z(x)$ consists of a deterministic first-order polynomial plus a random function described through a correlation function.

In this chapter we will show how we can generalize the concepts and methods of Chapter 3 to apply to the more general model of Equations (6.6) and (6.7).

6.3 Estimation with drift

Linear minimum-variance unbiased estimation can be readily extended to the linear model with variable mean and is known in geostatistics by the name *universal kriging* or *kriging with a drift*. The spatial function is given by the linear model

$$z(\mathbf{x}) = \sum_{k=1}^{p} f_k(\mathbf{x})\beta_k + \epsilon(\mathbf{x}), \tag{6.9}$$

where $f_1(\mathbf{x}), \ldots, f_p(\mathbf{x})$ are known functions of the spatial coordinates $\mathbf{x}$; $\beta_1, \ldots, \beta_p$ are deterministic but unknown coefficients; and $\epsilon(\mathbf{x})$ is a zero-mean random field.

It is assumed here that $\epsilon(\mathbf{x})$ in Equation (6.9) has a known covariance function, $R(\mathbf{x}, \mathbf{x}')$. We have not yet assumed any form of stationarity or discussed how to determine R from data; we will do so in other sections.

The problem is to obtain an estimate of z at $\mathbf{x}_0$ from measurements $z(\mathbf{x}_1), \ldots, z(\mathbf{x}_n)$. As in ordinary kriging, we require that:

1. The estimate be a linear function of available data $z(\mathbf{x}_1), z(\mathbf{x}_2), \ldots, z(\mathbf{x}_n)$, *i.e.*,

$$\hat{z}(\mathbf{x}_0) = \sum_{i=1}^{n} \lambda_i z(\mathbf{x}_i). \tag{6.10}$$

2. The coefficients $\lambda_1, \lambda_2, \ldots, \lambda_n$ be selected so that the estimate is unbiased, *i.e.*,

$$E[\hat{z}(\mathbf{x}_0) - z(\mathbf{x}_0)] = 0 \tag{6.11}$$

for any value of the unknown drift coefficients $\beta_1, \ldots, \beta_p$.

3. The estimation variance

$$E[(\hat{z}(\mathbf{x}_0) - z(\mathbf{x}_0))^2] \tag{6.12}$$

be as small as possible.

The unbiasedness condition (6.11) may be written

$$E\left[\sum_{i=1}^{n} \lambda_i z(\mathbf{x}_i) - z(\mathbf{x}_0)\right] = 0, \tag{6.13}$$

and, making use of (6.7), we get

$$\sum_{k=1}^{p}\left(\sum_{i=1}^{n} \lambda_i f_k(\mathbf{x}_i) - f_k(\mathbf{x}_0))\right)\beta_k = 0. \tag{6.14}$$

For this condition to hold for any values of $\beta_1, \ldots, \beta_p$, it is required that

$$\sum_{i=1}^{n} \lambda_i f_k(\mathbf{x}_i) = f_k(\mathbf{x}_0), \quad k = 1, \ldots, p. \tag{6.15}$$

This set of p constraints that must be satisfied by the λ coefficients is known among geostatisticians as the *universality conditions*, although "unbiasedness conditions" or even "elimination conditions" are more descriptive terms. Note that each unbiasedness condition has the effect of eliminating an unknown drift coefficient. (There are, of course, as many conditions as there are unknown drift coefficients.) Intuitively, we want to estimate $z(\mathbf{x}_0)$ from $z(x_1), \ldots, z(x_n)$ but we must first eliminate the unknown β coefficients that are in the way.

Example 6.2 *To illustrate the meaning of the unbiasedness conditions, assume that $z(\mathbf{x})$ is the piezometric head of a two-dimensional aquifer and $m(\mathbf{x})$ is a linear function. Then*

$$f_1 = 1, \quad f_2 = x_1, \quad f_3 = x_2, \tag{6.16}$$

where x_1 and x_2 are the two Cartesian coordinates of location $\mathbf{x}$. The unbiasedness conditions (6.15) become

$$\sum_{i=1}^{n} \lambda_i = 1, \quad \sum_{i=1}^{n} \lambda_i x_{i1} = x_{01}, \quad \sum_{i=1}^{n} \lambda_i x_{i2} = x_{02}. \tag{6.17}$$

We have assumed that the covariance function of $\epsilon(\mathbf{x})$ is $R(\mathbf{x}, \mathbf{x}')$. Making use of the unbiasedness condition, we find that the variance of the estimation error is

$$\begin{aligned} E[(\hat{z}(\mathbf{x}_0) - z(\mathbf{x}_0))^2] &= E\left[\left(\sum_{i=1}^{n} \lambda_i \epsilon(\mathbf{x}_i) - \epsilon(\mathbf{x}_0)\right)^2\right] \\ &= \sum_{i=1}^{n}\sum_{j=1}^{n} \lambda_i \lambda_j R(\mathbf{x}_i, \mathbf{x}_j) - 2\sum_{i=1}^{n} \lambda_i R(\mathbf{x}_i, \mathbf{x}_0) + R(\mathbf{x}_0, \mathbf{x}_0). \end{aligned} \tag{6.18}$$

Thus, the coefficients $\lambda_1, \lambda_2, \ldots, \lambda_n$ will be estimated by minimizing the expression of Equation (6.18) subject to the p linear constraints of Equation (6.15). Using the method of Lagrange multipliers, the coefficients must be determined from the following system:

$$\sum_{j=1}^{n} R(\mathbf{x}_i, \mathbf{x}_j)\lambda_j + \sum_{k=1}^{p} f_k(\mathbf{x}_i)\nu_k = R(\mathbf{x}_i, \mathbf{x}_0), \quad i = 1, \ldots, n \tag{6.19}$$

$$\sum_{i=1}^{n} f_k(\mathbf{x}_i)\lambda_i = f_k(\mathbf{x}_0), \quad k = 1, \ldots, p, \tag{6.20}$$

where $2\nu_1, \ldots, 2\nu_p$ are Lagrange multipliers. The kriging system consists of $n + p$ linear equations with $n + p$ unknowns.

Note that once the λ coefficients have been determined, they can be substituted in Equation (6.18) to determine the mean square estimation error. Actually, it is more computationally efficient to use the following expression for the mean square estimation error:

$$E[(\hat{z}(\mathbf{x}_0) - z(\mathbf{x}_0))^2] = -\sum_{k=1}^{p} f_k(\mathbf{x}_0)\nu_k - \sum_{i=1}^{n} \lambda_i R(\mathbf{x}_i, \mathbf{x}_0) + R(\mathbf{x}_0, \mathbf{x}_0). \quad (6.21)$$

Equations (6.19–6.21) comprise the key results of this section.

***Example 6.3** Consider the case with the linear trend given by Equation (6.16). The coefficients must be selected by solving the following system of $n+3$ linear equations with $n + 3$ unknowns $\lambda_1, \lambda_2, \ldots, \lambda_n, \nu_1, \nu_2, \nu_3$:*

$$\sum_{j=1}^{n} R(\mathbf{x}_i, \mathbf{x}_j)\lambda_j + \nu_1 + \nu_2 x_{i1} + \nu_3 x_{i2} = R(\mathbf{x}_i, \mathbf{x}_0), \quad i = 1, \ldots, n$$

$$\sum_{i=1}^{n} \lambda_i = 1$$

$$\sum_{i=1}^{n} x_{i1}\lambda_i = x_{01}$$

$$\sum_{i=1}^{n} x_{i2}\lambda_i = x_{02},$$

and the mean square error is

$$E[(\hat{z}(\mathbf{x}_0) - z(\mathbf{x}_0))^2] = -\nu_1 - x_{01}\nu_2 - x_{02}\nu_3 - \sum_{i=1}^{n} \lambda_i R(\mathbf{x}_i, \mathbf{x}_0) + R(\mathbf{x}_0, \mathbf{x}_0).$$

We have written the covariance function in the most general form to emphasize that, contrary to what is sometimes thought, the kriging equations do not really require stationarity or isotropy of any kind. Such assumptions are introduced in practice in order to make possible the selection of a covariance function from measurements. In the special case of stationary and isotropic $\epsilon(\mathbf{x})$, instead of $R(\mathbf{x}, \mathbf{x}')$, use $R(\|\mathbf{x} - \mathbf{x}'\|)$.

6.4 Generalized covariance function

Consider the linear model of Section 6.2. In kriging with a drift, the estimation error

$$\hat{z}(\mathbf{x}_0) - z(\mathbf{x}_0) = \sum_{i=1}^{n} \lambda_i z(\mathbf{x}_i) - z(\mathbf{x}_0) \quad (6.22)$$

does not depend on the values of the drift coefficients because of the unbiasedness conditions (*i.e.*, the λ coefficients must satisfy Equation (6.15)). Because it has this property, we say that the estimation error is an *authorized* increment. (We will explain later the origin of the term.)

What is special about an authorized increment? It has some important properties:

1. Its expected value is zero.
2. Its mean square value can be calculated from the covariance of the stochastic part of the linear model (see Equation (6.18)).
3. Furthermore, as we will soon demonstrate, the mean square error actually depends on a kernel of the covariance function, which is known as a *generalized* covariance function.

The concept of a generalized covariance function is actually quite simple: It is the part of the covariance function that determines the mean square error of an authorized increment. We are already familiar with the concept. In Chapter 3, we saw that with stationary functions

$$R(|\mathbf{x}-\mathbf{x}'|) = R(0) - \gamma(|\mathbf{x}-\mathbf{x}'|) \tag{6.23}$$

and for ordinary kriging we only need the variogram. Therefore, $-\gamma(|\mathbf{x}-\mathbf{x}'|)$, which is the part of the covariance function that matters, is a generalized covariance for kriging with a constant mean.

In the case of kriging with a variable mean, we may have several unbiasedness constraints. In simple terms, every time we add a new constraint we reduce the importance of the covariance function in estimation. The reason is that by adding another term in the deterministic part, we rely less on the stochastic part for the description of the spatial variability of $z(\mathbf{x})$. Thus, in an intuitive sense, we are allowed to use a simplified (or "generalized") version of the covariance function that works just as well as the actual covariance function. From the standpoint of practical estimation applications, this is good news because it is easier to infer from data the generalized covariance function than it is to infer the actual covariance function.

In general terms, we can view the covariance function as consisting of two parts,

$$R(\mathbf{x},\mathbf{x}') = K(\mathbf{x},\mathbf{x}') + C(\mathbf{x},\mathbf{x}'), \tag{6.24}$$

where $K(\mathbf{x},\mathbf{x}')$ is (by definition) the generalized covariance function and $C(\mathbf{x},\mathbf{x}')$ satisfies the condition

$$\sum_{i=1}^{n}\sum_{j=1}^{n}\lambda_i\lambda_j C(\mathbf{x}_i,\mathbf{x}_j) - 2\sum_{i=1}^{n}\lambda_i C(\mathbf{x}_i,\mathbf{x}_0) + C(\mathbf{x}_0,\ \mathbf{x}_0) = 0 \tag{6.25}$$

for any n and locations $\mathbf{x}_0, \mathbf{x}_1, \mathbf{x}_2, \ldots, \mathbf{x}_n$. Then, only $K(\mathbf{x}, \mathbf{x}')$ matters in kriging.

The reader is warned that interpretation of a generalized covariance function independently of the authorized increments that it refers to may be totally misleading. A generalized covariance function is not the actual covariance between two random variables; however, it can be used to calculate the variance of and the covariance between authorized increments. (Indeed, the term "authorized" can be interpreted as "authorized to use the generalized covariance function.") More on generalized covariance functions can be found in references [23, 99, 140].

Example 6.4 *Consider an intrinsic function $z(\mathbf{x})$ with generalized covariance function*

$$K(|\mathbf{x} - \mathbf{x}'|) = -|\mathbf{x} - \mathbf{x}'|.$$

Can we use it to find the mean square value of $z(\mathbf{x}) - z(\mathbf{x}')$?

Note that $z(\mathbf{x}) - z(\mathbf{x}')$ is not affected by the value of the mean and, consequently, is an authorized increment. Thus, we can go ahead to compute the mean square value of $z(\mathbf{x}) - z(\mathbf{x}')$ as if $z(x)$ had zero mean and covariance function $-|\mathbf{x} - \mathbf{x}'|$:

$$E[(z(\mathbf{x}) - z(\mathbf{x}')^2] = E[z(\mathbf{x})^2 + z(\mathbf{x}')^2 - 2z(\mathbf{x})z(\mathbf{x}')] = 2|\mathbf{x} - \mathbf{x}'|.$$

This may be rewritten as

$$\frac{1}{2}E[(z(\mathbf{x}) - z(\mathbf{x}'))^2] = |\mathbf{x} - \mathbf{x}'|,$$

i.e., the generalized covariance is effectively the same as our familiar linear variogram. However, $z(\mathbf{x})$ or $\frac{z(\mathbf{x})+z(\mathbf{x}')}{2}$ are not authorized increments. Thus, if we use the generalized covariance function as if it were the true covariance function, we obtain absurd results, such as zero variability,

$$E[(z(\mathbf{x}))^2] = 0,$$

or, even worse, negative mean square error,

$$E\left[\left(\frac{z(\mathbf{x}) + z(\mathbf{x}')}{2}\right)^2\right] = -\frac{1}{2}|\mathbf{x} - \mathbf{x}'|.$$

Example 6.5 *Consider a random function $z(x)$ defined on one dimension with a linear drift. Take the increment*

$$y = z(x - a) - 2z(x) + z(x + a). \tag{6.26}$$

If a linear trend, $b_1 + b_2 x$, is added to $z(x)$, then

$$\begin{aligned} y = b_1 + (x - a)b_2 + z(x - a) - 2(b_1 + xb_2 + z(x)) + b_1 + (x + a)b_2 \\ + z(x + a) = z(x - a) - 2z(x) + z(x + a), \end{aligned} \tag{6.27}$$

which is exactly the same as y in Equation (6.26). This means that y is an authorized increment.

6.5 Illustration of the GCF concept

This section illustrates through a detailed example some of the points made in the previous section. It is heavy in algebra and you may prefer to skip this section at first reading.

We will consider a spatial function z defined on two dimensions (x, y) that is modeled as the sum of a linear trend and a zero-mean stochastic process:

$$z(x, y) = \beta_1 + \beta_2 x + \beta_3 y + \epsilon(x, y), \tag{6.28}$$

where x and y are spatial coordinates; β_1, β_2, and β_3 are unknown drift coefficients; and ϵ is a zero-mean stochastic process with covariance function, for example,

$$R(h) = \sigma^2 \exp\left(-\frac{h}{l}\right), \tag{6.29}$$

where $h = \sqrt{(x - x')^2 + (y - y')^2}$ is distance and σ^2 and l are parameters.

Given these observations (which we will denote by $z_1, z_2, \ldots, z_n$) we want to estimate z_0, the value of z at location (x_0, y_0). Limiting our attention to linear estimators,

$$\hat{z}_0 = \sum_{i=1}^{n} \lambda_i z_i, \tag{6.30}$$

our task is to select $\lambda_1, \lambda_2, \ldots, \lambda_n$. We will enforce the unbiasedness condition, *i.e.*, that the expected value of the estimation error should be zero,

$$E\left[\sum_{i=1}^{n} \lambda_i z_i - z_0\right] = 0. \tag{6.31}$$

Using Equation 6.28, we have that

$$E\left[\sum_{i=1}^{n} \lambda_i (\beta_1 + \beta_2 x_i + \beta_3 y_i + \epsilon(x_i, y_i)) - (\beta_1 + \beta_2 x_0 + \beta_3 y_0 + \epsilon(x_0, y_0))\right] = 0 \tag{6.32}$$

or, using the fact that ϵ is a zero-mean process and rearranging terms, we get

$$\left(\sum_{i=1}^{n} \lambda_i - 1\right)\beta_1 + \left(\sum_{i=1}^{n} \lambda_i x_i - x_0\right)\beta_2 + \left(\sum_{i=1}^{n} \lambda_i y_i - y_0\right)\beta_3 = 0. \tag{6.33}$$

The only way to guarantee that this expression will be zero for any value of β_1, β_2, and β_3 is to enforce the unbiasedness constraints:

$$\sum_{i=1}^{n} \lambda_i = 1 \tag{6.34}$$

$$\sum_{i=1}^{n} \lambda_i x_i = x_0 \tag{6.35}$$

$$\sum_{i=1}^{n} \lambda_i y_i = y_0. \tag{6.36}$$

Next, we compute the mean square error

$$MSE = E\left[\left(\sum_{i=1}^{n} \lambda_i z_i - z_0\right)^2\right]. \tag{6.37}$$

Making use of the unbiasedness constraints, we obtain

$$\begin{aligned} MSE &= E\left[\left(\sum_{i=1}^{n} \lambda_i (z_i - E[z_i]) - (z_0 - E[z_0])\right)^2\right] \\ &= \sum_{i=1}^{n}\sum_{j=1}^{n} \lambda_i \lambda_j R(h_{ij}) - 2\sum_{i=1}^{n} \lambda_i R(h_{i0}) + R(0), \end{aligned} \tag{6.38}$$

where h_{ij} is the distance between the locations of measurements i and j and h_{i0} is the distance between the locations of measurement i and the unknown. Using the method of Lagrange multipliers, the solution that minimizes the mean square error subject to unbiasedness constraints is given by solving the system of $n + 3$ equations with $n + 3$ unknowns ($\lambda_1, \lambda_2, \ldots, \lambda_n, \nu_1, \nu_2, \nu_3$):

$$\sum_{j=1}^{n} \lambda_j R(h_{ij}) + \nu_1 + \nu_2 x_i + \nu_3 y_i = R(h_{i0}) \quad \text{for } i = 1, 2, \ldots, n \tag{6.39}$$

combined with Equations (6.34), (6.35), and (6.36).

We can now verify that, due to the unbiasedness constraint, the MSE is the same whether one uses $R(h)$ or $R(h) + a_0 + a_2 h^2$, where a_0 and a_2 are arbitrary coefficients,

$$\begin{aligned} MSE &= \sum_{i=1}^{n}\sum_{j=1}^{n} \lambda_i \lambda_j \left[R(h_{ij}) + a_0 + a_2 h_{ij}^2\right] \\ &\quad - 2\sum_{i=1}^{n} \lambda_i \left[R(h_{i0}) + a_0 + a_2 h_{i0}^2\right] + R(0) + a_0 \\ &= \sum_{i=1}^{n}\sum_{j=1}^{n} \lambda_i \lambda_j R(h_{ij}) - 2\sum_{i=1}^{n} \lambda_i R(h_{i0}) + R(0). \end{aligned} \tag{6.40}$$

The proof follows:

$$\sum_{i=1}^{n}\sum_{j=1}^{n}\lambda_i\lambda_j a_0 - 2\sum_{i=1}^{n}\lambda_i a_0 + a_0$$
$$= \left[\left(\sum_{i=1}^{n}\lambda_i\right)\left(\sum_{j=1}^{n}\lambda_j\right) - 2\left(\sum_{i=1}^{n}\lambda_i\right) + 1\right] a_0 = 0$$

because the sum of the λ coefficients is 1. Furthermore,

$$\sum_{i=1}^{n}\sum_{j=1}^{n}\lambda_i\lambda_j a_2 h_{ij}^2 - 2\sum_{i=1}^{n}\lambda_i a_2 h_{i0}^2$$
$$= \left[\sum_{i=1}^{n}\sum_{j=1}^{n}\lambda_i\lambda_j((x_i - x_j)^2 + (y_i - y_j)^2)\right.$$
$$\left. - 2\sum_{i=1}^{n}\lambda_i((x_i - x_0)^2 + (y_i - y_0)^2)\right] a_2$$
$$= \left[\sum_{i=1}^{n}\sum_{j=1}^{n}\lambda_i\lambda_j(x_i^2 + x_j^2 - 2x_i x_j + y_i^2 + y_j^2 - 2y_i y_j)\right.$$
$$\left. - 2\sum_{i=1}^{n}\lambda_i(x_i^2 + x_0^2 - 2x_i x_0 + y_i^2 + y_0^2 - 2y_i y_0)\right] a_2 = 0. \quad (6.41)$$

We will show that the bracketed expression is zero. This expression can be factored into the form

$$\sum_{i=1}^{n}\lambda_i(x_i^2 + y_i^2)\left(\sum_{j=1}^{n}\lambda_j\right) + \left(\sum_{i=1}^{n}\lambda_i\right)\sum_{j=1}^{n}\lambda_j(x_j^2 + y_j^2)$$
$$- 2\left(\sum_{i=1}^{n}\lambda_i x_i\right)\left(\sum_{j=1}^{n}\lambda_j x_j\right) - 2\left(\sum_{i=1}^{n}\lambda_i y_i\right)\left(\sum_{j=1}^{n}\lambda_j y_j\right)$$
$$- 2\sum_{i=1}^{n}\lambda_i(x_i^2 + y_i^2) + 4\left(\sum_{i=1}^{n}\lambda_i x_i\right)x_0 + 4\left(\sum_{i=1}^{n}\lambda_i y_i\right)y_0$$
$$- 2\left(\sum_{i=1}^{n}\lambda_i\right)(x_0^2 + y_0^2) = 2\sum_{i=1}^{n}\lambda_i(x_i^2 + y_i^2) - 2x_0^2 - 2y_0^2$$
$$- 2\sum_{i=1}^{n}\lambda_i(x_i^2 + y_i^2) + 4x_0^2 + 4y_0^2 - 2(x_0^2 + y_0^2) = 0. \quad (6.42)$$

Therefore, adding the term $a_1 + a_2 h^2$ to the covariance funtion has no effect on the mean square error or the associated kriging equations. Consequently: (*a*) These terms cannot be identified from data and (*b*) including these terms in the model results in "overparameterization."

6.6 Polynomial GCF

This section presents an example that also constitutes a useful model.

We will focus on one particular isotropic generalized covariance function, known as the *polynomial generalized covariance function* (or PGCF). We discuss it in some detail because it is useful in applications and can serve to clarify a number of points associated with kriging with a drift.

Matheron [96] extended the idea of intrinsic functions to intrinsic functions of higher order. Let us start by saying that:

The drift of an intrinsic function of k-th order is a polynomial of order k.

Example 6.6 *Consider the one-dimensional case. Referring to the linear model, Equation 6.7, the zeroth-order intrinsic function corresponds to $p = 1$ and $f_1(x) = 1$. Thus, what we called in Chapter 3 "an intrinsic function" becomes in this chapter "a 0-order intrinsic function." The first-order intrinsic function corresponds to $p = 2$, $f_1 = 1$, and $f_2 = x$. The second-order intrinsic function corresponds to $p = 3$, $f_1 = 1$, $f_2 = x$, and $f_3 = x^2$.*

Thus, the base functions of the mean of an intrinsic function of order k are monomials of order k or less.

We are now ready to present a model useful in estimation. In all cases, the model is

$$z(\mathbf{x}) = m(\mathbf{x}) + \epsilon(\mathbf{x}), \tag{6.43}$$

where $m(\mathbf{x})$ is a k-th-order polynomial and $\epsilon(\mathbf{x})$ is a zero-mean stationary process with isotropic covariance function $R(h)$, $h = |\mathbf{x} - \mathbf{x}'|$. (We will soon see that the assumption of stationarity can be relaxed.) To fix ideas, assume that $z(\mathbf{x})$ is defined in the two-dimensional space, $\mathbf{x} = \binom{x_1}{x_2}$. Then

$$k = 0, \quad m(x_1, x_2) = 1.\beta_1, \tag{6.44}$$

$$k = 1, \quad m(x_1, x_2) = 1.\beta_1 + x_1\beta_2 + x_2\beta_3, \tag{6.45}$$

$$k = 2, \quad m(x_1, x_2) = 1.\beta_1 + x_1\beta_2 + x_2\beta_3 + x_1^2\beta_4 + x_2^2\beta_5 + x_1x_2\beta_6, \tag{6.46}$$

and so on.

The generalized covariance function also includes the nugget-effect term $C_0\delta(h)$, where

$$\delta(h) = \begin{cases} 0, & \text{if } h > 0 \\ 1, & \text{if } h = 0 \end{cases}. \tag{6.47}$$

Thus,

$$K(h) = C_0\delta(h) + \theta_1 h, \tag{6.48}$$

where $K(h)$ is the polynomial generalized covariance function and the two parameters C_0 and θ_1 must satisfy

$$C_0 \geq 0 \quad \text{and} \quad \theta_1 \leq 0. \tag{6.49}$$

For $k = 1$, according to reference [38],

$$K(h) = C_0\delta(h) + \theta_1 h + \theta_3 h^3, \tag{6.50}$$

where

$$C_0 \geq 0, \quad \theta_1 \leq 0, \quad \text{and} \quad \theta_3 \geq 0, \tag{6.51}$$

and, for $k = 2$,

$$K(h) = C_0\delta(h) + \theta_1 h + \theta_3 h^3 + \theta_5 h^5, \tag{6.52}$$

where

$$C_0 \geq 0, \quad \theta_1 \leq 0, \quad \text{and} \quad \theta_5 \leq 0$$

and

$$\theta_3 \geq \begin{cases} -(10/3)[\theta_1\theta_5]^{1/2}, & \text{in two dimensions} \\ -[10\theta_1\theta_5]^{1/2}, & \text{in three dimensions} \end{cases} \tag{6.53}$$

Application of this model is discussed in reference [77]. Extension to *anisotropic* cases is straightforward by making use of the geoanisotropic model. The basic idea is that, after a combination of rotations and stretchings of the spatial coordinates, the generalized covariance function becomes a function of distance.

6.7 Stationary-increment processes

In the kriging applications that we have seen, we have dealt only with authorized increments. We may expand the concept of stationarity accordingly. This leads us to the class of random functions with *stationary increments*. An operational definition of such functions is:

> A random field $z(x)$ is stationary-increment with generalized covariance function $K(\mathbf{x} - \mathbf{x}')$ if the mean of any authorized increments is zero and the covariance of any two authorized increments can be calculated using K as if $z(x)$ were stationary with zero mean and covariance function K.

An example of a stationary-increment process is an intrinsic function of order k with a polynomial generalized covariance function.

One cannot overemphasize that stationary-increment random fields are defined and can be interpreted only in terms of authorized increments, just like our familiar intrinsic random field. Obviously, the stationary-increment process is a generalization of the intrinsic function of Chapters 3 and the generalized covariance function is a generalization of the variogram.

We have seen that in the case of intrinsic functions (constant mean) the continuity and differentiability of the realizations depend on the behavior of the variogram at the origin (zero distance). In the case of functions with variable (but continuous and differentiable) mean the following properties hold:

1. If the drift is continuous, the realizations are continuous when the generalized covariance is continuous at the origin.
2. If the drift is differentiable, the realizations are differentiable when the generalized covariance is twice differentiable (*i.e.*, has parabolic behavior) at the origin.

6.8 Splines

The scope of geostatistics has gradually expanded to include applications that traditionally have been addressed using spline interpolation methods. Moreover, the relation of kriging or Bayesian estimation to interpolating splines has been explored in references [30, 47, 78, 97, 142] and others.

Consider, for example, kriging with a first-order polynomial with unknown coefficients for a mean and a generalized covariance function given by

$$R(\mathbf{x} - \mathbf{x}') = \theta \|\mathbf{x} - \mathbf{x}'\|^3, \tag{6.54}$$

where θ is a positive parameter. The estimate reproduces the data and is smooth, possessing continuous first and second derivatives everywhere. This case corresponds to a form of cubic-spline interpolation.

In two dimensions, a useful model is one in which the mean is a first-order polynomial with unknown coefficients and the generalized covariance function is

$$R(\mathbf{x} - \mathbf{x}') = \theta \|\mathbf{x} - \mathbf{x}'\|^2 \log \|\mathbf{x} - \mathbf{x}'\|. \tag{6.55}$$

This case corresponds to interpolation through "thin-plate" (in two-dimensional applications) splines, which gives the simplest smooth curve that reproduces the data. The estimate does not have well-defined second derivatives at observation points but has continuous first derivatives everywhere and is flatter and supposedly more "elegant" than the estimate in the cubic-spline interpolation method. Figure 6.4 illustrates some of these points for an easy-to-plot one-dimensional case.

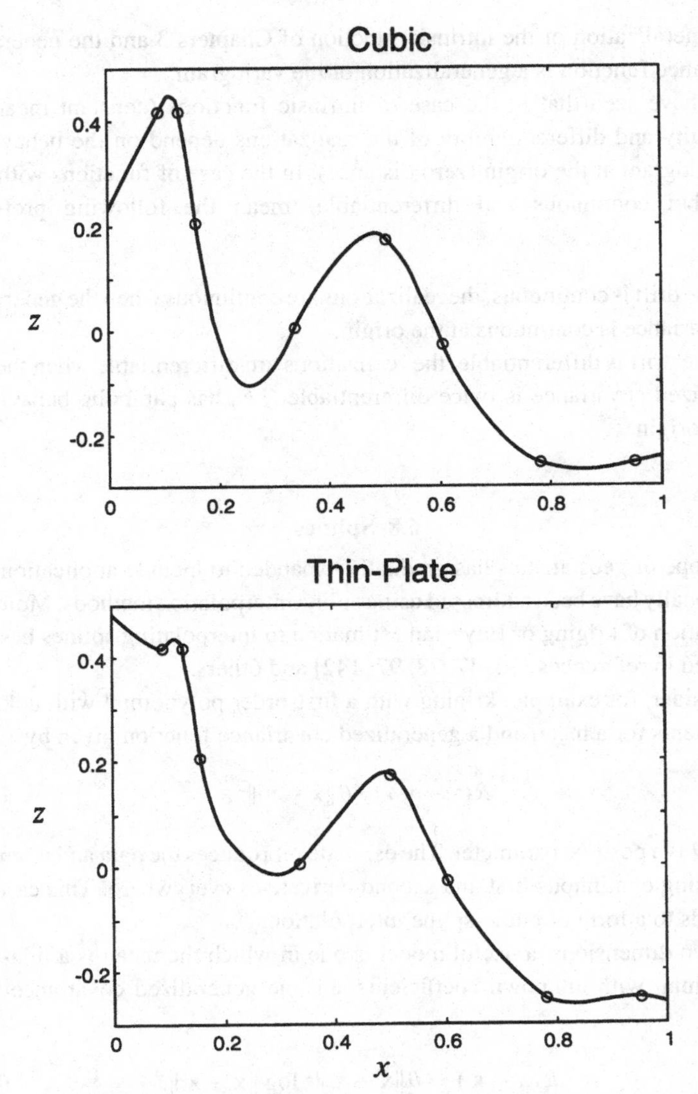

Figure 6.4 Interpolation through data (shown as ○) using cubic and thin-plate splines.

6.9 Validation tests

We can now return to the problem of data analysis with an enlarged repertory of models, since we can deal with variable-mean models. As usual, however, model selection is an iterative (trial and error) process in which a model is proposed, calibrated, checked, modified, and so on until a satisfactory one

Suggestions for Developing the Model:
1. Adopt variable-mean model only if it fits the data significantly better than the constant-fit model or if there is other justification.
2. Obtain preliminary estimate of variogram from the experimental variogram of the detrended data.
3. Calibrate variogram (estimate parameters) and validate model using residuals.

Figure 6.5 How to proceed in developing a model with variable mean.

is found. Figure 6.5 gives some useful hints on how to proceed when the mean is variable. We will elaborate on these points in the remainder of this chapter.

We will start by discussing the method of residuals as a way of validating or testing a model. Suppose that we have somehow developed a model, such as a first-order intrinsic one with a nugget and a linear term. Before the proposed model is used for estimation, its consistency with the data must be ascertained through what we called model validation. The methods that we saw in Chapter 3 are generally applicable to the case of variable drift. Here, we will focus on the same relatively simple tests involving orthonormal residuals that we saw in Chapter 4.

Consider the following procedure: The n measurements may be shuffled to obtain, randomly, a fresh order in the sample. Then, calculate the kriging estimate of z at the $(p+1)$-th point, x_{p+1}, given only the first p measurement, $x_1, x_2, \ldots, x_p$. (p is the number of drift coefficients.) Calculate the actual error $z(x_{p+1}) - \hat{z}_{p+1}$ and normalize it by the standard error σ_{p+1}. The normalized error is called ε_{p+1}. For the k-th ($k = p+1, p+2, \ldots, n$) measurement location we estimate through kriging the value of z using only the previous

$k-1$ measurements, and we normalize by the standard error. Thus,

$$\varepsilon_k = \frac{z(x_k) - \hat{z}_k}{\sigma_k}, \quad k = p+1, \ldots, n. \tag{6.56}$$

We thereby obtain $n-p$ residuals, which, as one can verify, have the orthonormality property:

$$E[\varepsilon_k \varepsilon_\ell] = \begin{cases} 1, & \text{if } k = \ell \\ 0, & \text{if } k \neq \ell \end{cases} \quad k, \ell = p+1, \ldots, n. \tag{6.57}$$

The proof follows the same lines as for the intrinsic case (see Chapter 3). Now, consider the statistics

$$Q_1 = \frac{1}{n-p} \sum_{k=p+1}^{n} \varepsilon_k \tag{6.58}$$

$$Q_2 = \frac{1}{n-p} \sum_{k=p+1}^{n} \varepsilon_k^2. \tag{6.59}$$

If the model is consistent with the data, the first number must be near zero, while the second must be near one. The sampling distribution of Q_1 and Q_2 can be obtained easily:

$$E[Q_1] = 0 \tag{6.60}$$

$$E\left[Q_1^2\right] = \frac{1}{n-p}. \tag{6.61}$$

In applications it is usually sufficient to assume that Q_1 is normally distributed so that one can use the following rule:

Reject the model if

$$|Q_1| > 2/\sqrt{n-p}. \tag{6.62}$$

When using this rule, the probability of rejecting the model even though the model is good is about 5%.

Regarding Q_2, we see that

$$E[Q_2] = 1. \tag{6.63}$$

The variance of Q_2 can be easily calculated for Gaussian z:

$$E[(Q_2 - 1)^2] = \frac{2}{n-p}. \tag{6.64}$$

Furthermore, still under the Gaussian assumption, $(n-p)Q_2$ is chi-square distributed with $(n-p)$ degrees of freedom, which leads to the rule:

Reject the model if

$$Q_2 > U \quad \text{or} \quad Q_2 < L, \tag{6.65}$$

where the values of U and L can be found from a table in a handbook of statistical methods. Actually, for $n > 40$, approximating the distribution of Q_2 by a Gaussian is adequate for this application so that one can use the following rule:

Reject the model if

$$|Q_2 - 1| > 2.8/\sqrt{n-p}. \tag{6.66}$$

The ε residuals can also be used to test the hypothesis that they were generated from a Gaussian distribution and to locate outliers. Also, one may perform a variogram analysis and test the hypothesis that the experimental variogram of $\varepsilon_{p+1}, \ldots, \varepsilon_n$ is indeed a pure nugget effect.

6.10 Parameter estimation

We assume that the model is known and the problem is to estimate the parameters $\theta_1, \ldots, \theta_m$ of the generalized covariance function.

We can use the same method that we used in Chapter 3. Simply speaking, when the correct parameters are used and many measurements are available, the sample value of the statistic Q_2 should be near 1. We are now faced with the inverse problem: The parameters are not known, but we can calculate the value of Q_2 for any set of parameters. It is reasonable to select the parameters that make Q_2 equal to 1:

$$Q_2 = \frac{1}{n-p} \sum_{k=p+1}^{n} \frac{\delta_k^2}{\sigma_k^2} = 1. \tag{6.67}$$

If there is only one unknown parameter, it can be estimated from this criterion. Otherwise, there may be more than one set of parameter estimates satisfying this criterion. Good parameters should result in small estimation errors. Thus, it is reasonable to add the criterion that the parameter estimates should minimize, in some sense, the square kriging errors $\delta_{1+p}^2, \ldots, \delta_n^2$. An alternative approach is to use, instead of the square kriging error, δ_k^2, its average value, σ_k^2. Because we have imposed requirement (6.67), σ_k^2 is a good measure of δ_k^2 and is also less affected by randomness. Thus, we may introduce the requirement

$$\min \frac{1}{n-p} \sum_{k=p+1}^{n} \ell n(\sigma_k^2), \tag{6.68}$$

or some other measure of overall accuracy, and obtain the following parameter-estimation method:

Estimate the parameters that minimize the expression of (6.68) subject to the constraint (6.67).

Actually, solutions to this optimization problem are easily obtained for the case of the polynomial generalized covariance function model. We will discuss how estimates can be obtained in the most commonly encountered cases of one and two parameters:

1. One parameter. The procedure in the case of only one parameter is the following: Set the value of the parameter equal to one. The parameter is then equal to the computed value of $\frac{1}{n-p}\sum_{k=p+1}^{n}\varepsilon_k^2$. No iterations are needed.
2. Two parameters. The main objective is to determine the ratio of the two parameters θ_1 and θ_2. We may set θ_2 equal to some reasonable value and vary θ_1. For each value of θ_1:
   - Using the current estimates of the parameters, calculate kriging errors δ_k and kriging variances σ_k^2.
   - Calculate $Q_2 = \frac{1}{n-p}\sum_{k=p+1}^{n}\delta_k^2/\sigma_k^2$.
   - Calculate $C(\theta_1/\theta_2) = \frac{1}{n-p}\sum_{k=p+1}^{n}\ln(\sigma_k^2) + \ln(Q_2)$.
   - Select the ratio θ_1/θ_2 for which C is minimum. Then adjust parameters by multiplying them by the value of Q_2 which corresponds to the minimum.

One might raise the following question: What is the accuracy of parameter estimates? Let us start with the case of only one parameter. The parameter estimate is the value that makes the value of Q_2 equal to 1. In fact, Q_2 is a sample value of a random variable whose mean value is 1. The variance, for Gaussian z and large sample, is $2/(n-p)$. By the method we use to estimate the parameters, one can conclude that the variance of estimation of the parameter is proportional to $2/(n-p)$. Thus, if the sample is small, the parameter estimate obtained from the data may not be reliable. In the case of many parameters, an error analysis is less straightforward. However, one can reason that the accuracy will depend on the curvature of C. The more curved or peaked that $C(\theta_1/\theta_2)$ is, the more accurate are the results.

6.11 On model selection

We have seen that kriging with a drift is not significantly more complicated than kriging with a constant mean. Even parameter estimation is a relatively straightforward procedure once we select a model and a fitting criterion. The solution to these problems follows well-defined rules and can be automated. We now start considering how to choose the model, which in our case means the form of the drift and the generalized covariance function. Unlike kriging, which is a straightforward mathematical problem given the model and its parameters, this part involves induction, the development of a model from data and other information, and, consequently, it is a different game.

As a matter of principle, one cannot "deduce" or "prove" a model on the basis of limited information, such as a finite sample. This is true in all sciences that deal with empirical data. "Inducing" a model means that a model is proposed and found to be consistent with available evidence. Model development is a trial and error procedure in which a series of assumptions are tentatively proposed and then tested. However, one should not think that induction does not have its own rules. What follows is a discussion of the most important of these rules, presented from a practical standpoint.

1. Choose the simplest model consistent with past experience with similar data. For example, it is often reasonable to start with the assumption that the regionalized variable is intrinsic and isotropic and to determine its experimental variogram. This model should, however, be put to the test using validation tests, such as those described in Chapter 3. If model inadequacies are detected, one should relax the assumption of constant mean or of isotropy. One should be careful to do so gradually so that the model can stay as simple as possible. For example, one should add complexity by introducing either anisotropy or variable mean, not both at the same time.

   The drift model is selected on the basis of past experience, prior information, the data, and the objectives of modeling. In practice, variable-mean models should be used only when there is strong evidence from prior information and the data. This topic has been discussed in reference [137].

   Regarding covariance functions, a useful rule of thumb is to avoid covariance models that involve more than two unknown parameters. (Of course, exceptions can be made in special cases.) Models with many parameters do not necessarily yield better predictions. If anything, the opposite is the case.
2. We should always strive to pick the right model, which means the one that most accurately represents our prior information and what we have learned from the new measurements. However, if we are unsure about what is the actual spatial variability, it is preferable to choose a model that does not prejudice the solution by much. For example, one of the appealing aspects of the intrinsic hypothesis is that it does not predetermine the solution but lets the data speak for themselves. As we have already pointed out, the intrinsic model is a prudent choice when little information is available. As more measurements are obtained, the actual structure starts to emerge, at which point we may want to adopt a different assumption.
3. Keep in mind that the model is usually developed to solve a specific prediction problem. Thus, it is of little practical usefulness to keep searching for the "true" model when one model has already been found that adequately

represents what is known about the spatial function of interest and seems to fit the data adequately.

In some cases, one may come up with many models which appear equally plausible given the available information. Instead of agonizing about which one to choose, one should calculate the predictions using each of these models and then compare the results. In many cases, one finds that the results are practically the same, and hence there would be no point in fretting about model uncertainty. However, if results differ, a more logical and honest approach would be to present the results of all plausible models.

4. Structural analysis does not rely solely on available observations. Sometimes observations are not sufficient to figure out some important feature of the covariance function (or variogram). For example, the exact behavior near the origin may be hard to determine on the basis of point measurements, and so the analyst must exercise his judgement. Ultimately, a practitioner would choose the model of spatial structure that has the most important features that the true function is perceived to possess.

   Most engineers and geophysicists seem comfortable with the idea that their knowledge and experience not only can but also should be used. Specific measurements or "hard data" are not the only source of objective information. Measurements are affected by error and must be interpreted very carefully in the context of a conceptual model of the actual process. To be sure, hard data contain valuable information—which we labor to extract—but so does the hydrogeologists's understanding of the geological processes and the mechanisms of flow and transport.

5. The importance of prior information is inversely proportional to the size of the data. Actually, with the methods presented in this book, we usually assume the availability of a set of measurements of reasonable size that contains most information about the variable of interest. However, you realize that to try to determine a variogram from, say, only two or three measurements is futile. If the number of observations is so small, it is preferable to use prior information to establish a variogram. For example, if the variable is log-conductivity (natural logarithm of conductivity), based on the information about the type of the formation one should be able to assign a value to the variance. For example, assume we know that the formation is sandy. Because sand varies widely in coarseness, it is expected that the conductivity k may vary over four orders of magnitude. The log-conductivity ($\ln k$) varies by $\ln 10^4 = 9.2$. An estimate of the $\ln k$ variance is $\frac{1}{4}$ of this range squared: $(\frac{9.2}{4})^2 \simeq 5$. It is obvious that in this case, the variance represents what little is known about the formation and is not to be interpreted as a range of actual spatial variability at the specific formation.

6.12 Data detrending

We turn our attention now to the problem of obtaining a preliminary estimate of the covariance. In ordinary kriging (with constant mean) most practitioners graphically fit an equation to the experimental variogram. The same approach has been extended to kriging with variable mean by using the detrended data, *i.e.*, the original data from which the fitted drift has been subtracted. This approach has been criticized [5] because the variogram of the detrended data is different from the variogram of the original stochastic process and depends on the method of detrending. It is well known [129] that the presence of a drift distorts the experimental variogram.

It turns out that although the detrended data has a different covariance function (or variogram) from the original process, *the original and the detrended data have the same GCF* [83]. The practical significance of this result is that:

> The experimental covariance (or variogram) of the detrended data can be used to estimate the generalized (not the ordinary) covariance function. The simplest and most computationally efficient approach is to detrend using ordinary least squares.

Since the original and the detrended data share the same generalized covariance, one may ask what are the advantages of using the detrended data instead of the original data in experimental variogram analysis. The answer is that, when the original data are used, the trend swamps the experimental variogram, making the job of inferring the generalized covariance function even more difficult. Detrending is helpful because it removes much of the redundant (drift-related) variability in the experimental variogram, thereby revealing the generalized covariance function.

6.13 An application

Following reference [83], we will analyze hydraulic head data from the Jordan aquifer in Iowa [64]. The data are given in Table 2.1 and the location of the measurements is shown on Figure 2.16. As is often the case with hydraulic head in deep aquifers at a regional scale, the data indicate an approximate linear drift. The nodes of the experimental variogram of the data are shown on Figure 6.6. Based on statistical arguments as well as additional hydrogeologic information not part of the data set, it was deemed appropriate to try a variable mean model with

$$m(x_1, x_2) = \beta_1 + \beta_2 x_1 + \beta_3 x_2 \qquad (6.69)$$

where x_1 and x_2 are spatial coordinates and β_1, β_2, and β_3 are deterministic but unknown coefficients.

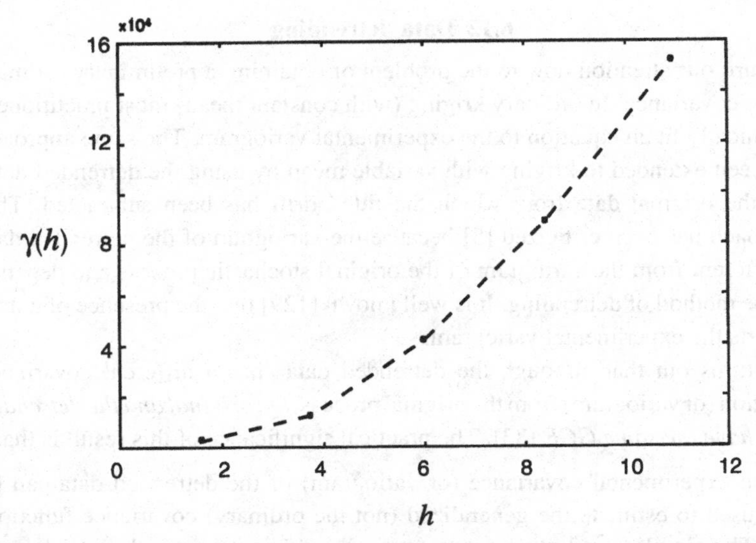

Figure 6.6 Experimental (semi)variogram of original data. (Adapted after [83].)

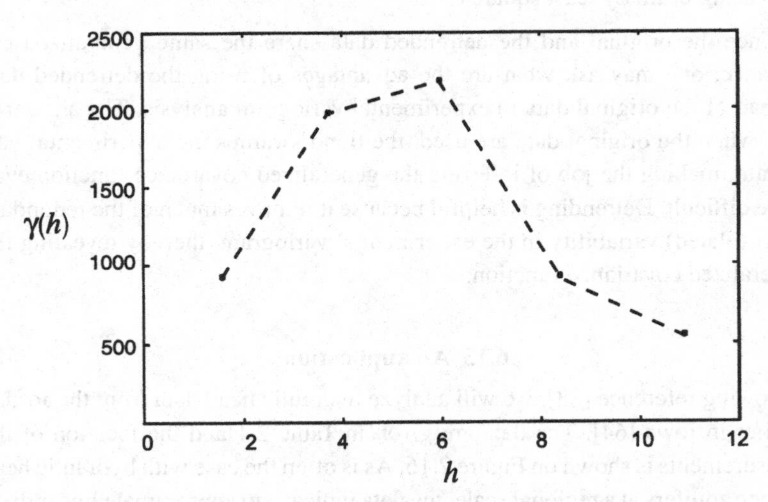

Figure 6.7 Experimental (semi)variogram of detrended data. (Adapted after [83].)

The experimental variogram of the original data (Figure 6.6) is not very helpful because it suggests a quadratic power model. However, an h^2 term is redundant in kriging with a linear drift (Equation 6.69). The problem is that the drift has swamped the variability of the stochastic part in the experimental variogram. For this reason, the data were detrended and the variogram was plotted in Figure 6.7. It appears that an exponential variogram might be appropriate.

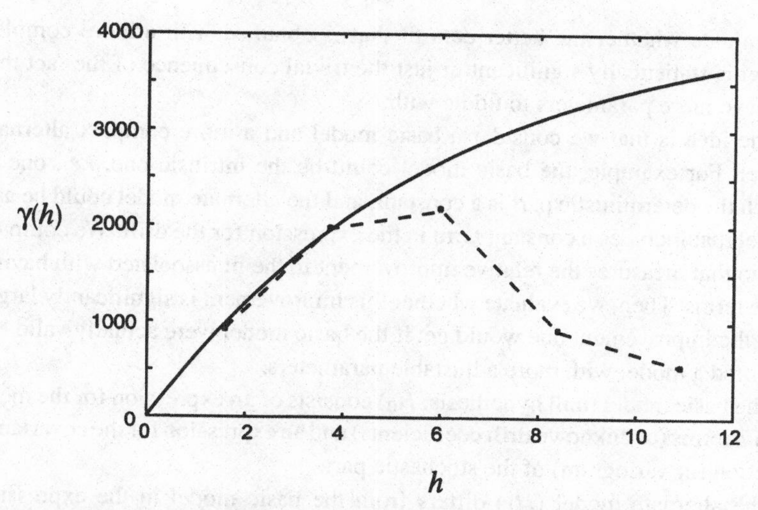

Figure 6.8 Experimental variogram and fitted equation (exponential GCF) using detrended data. (Adapted after [83].)

This corresponds to the following generalized covariance function:

$$K(h) = \theta_1 \exp(-h/\theta_2), \tag{6.70}$$

where h is separation distance and θ_1 and θ_2 are positive parameters. The parameters can be fitted graphically from the experimental variogram, but the procedure is somewhat subjective.

Another approach is to select the θ_2 that minimizes the prediction error and to select θ_1 from the mean square error of estimation in the method of orthonormal residuals. This method leads to estimates of $\hat{\theta}_1 = 4228$ and $\hat{\theta}_2 = 6$. Values of θ_2 in the range 5–7 provide essentially equally good fits. On Figure 6.8, the continuous line represents the fitted model and is shown to be in agreement with the experimental variogram. Note that the apparent discrepancy between the model and the experimental variogram at large lags is of no concern for two reasons: First, the sampling error associated with the experimental variogram at large lags is so large that it would be inappropriate to modify the model to obtain better reproduction of the experimental variogram at large distances. Second, variograms that differ by a quadratic function are practically the same for purposes of estimation when a linear drift is included in the model.

6.14 Variance-ratio test

One of the questions often raised in practice is whether to add terms to the drift (deterministic part) of a geostatistical model. Here, we describe a practical way to evaluate whether the better data fit that is obtained with a more complex model is statistically significant or just the trivial consequence of the fact that we have more parameters to fiddle with.

The idea is that we consider a basic model and a more complex alternate model. For example, the basic model could be the intrinsic one, *i.e.*, one in which the deterministic part is a constant, and the alternate model could be any model that includes a constant term in the expression for the drift. We compute a ratio that measures the relative improvement in the fit associated with having more terms. Then, we evaluate whether this improvement is significantly larger than the improvement one would get if the basic model were actually valid but we fitted a model with more adjustable parameters.

The basic model (null hypothesis, H_0) consists of an expression for the mean with p terms (or unknown drift coefficients) and an expression for the covariance function (or variogram) of the stochastic part.

The alternate model (H_1) differs from the basic model in the expression used for the mean. The alternate model has $p + q$ terms (drift coefficients). The important requirement is that the alternate model is more general than the basic model. That is, the basic model can be reduced from the alternate model by setting q different linear combinations of drift coefficients to zero. For illustration, consider the following examples:

Example 6.7 *H_0: $m(x) = \beta_1$, H_1: $m(x) = \beta_1' + \beta_2' x + \beta_3' x^2$. In this case, $p = 1$, $q = 2$, and the alternate model is more general because it can be reduced to the basic model if we set $\beta_2' = 0$ and $\beta_3' = 0$.*

Example 6.8 *H_0: $m(x) = \beta_1 + \beta_2(x + x^2)$, H_1: $m(x) = \beta_1' + \beta_2' x + \beta_3' x^2$. In this case, $p = 2$, $q = 1$, and the alternate model is more general because it can be reduced to the basic model if we set $\beta_2' - \beta_3' = 0$.*

We emphasize that the same covariance function that is found for the basic model must be used without modification for the alternate model.

We start with our basic model H_0, but we are considering whether we should reject it in favor of the alternate model H_1. If we look at an overall measure of how well the model reproduces the data after we fit its parameters to the data, we always find that the more general model gives a better fit, for the simple reason that the more general model has more parameters to adjust. This is not necessarily an indication that the alternate model is better for prediction purposes. We should reject our simpler model in favor of the more complicated one only if the fit is significantly better.

We will now construct a test that measures the fit for each model and addresses the issue of the significance in their difference. We will consider the following weighted sum of squares (*WSS*), which measures the agreement between model and data:

$$WSS_0 = \sum_{i=p+1}^{n} \varepsilon_i^2, \ H_0, \tag{6.71}$$

where ε are the orthonormal residuals computed for the basic model.

Then, we compute the same quantity using the orthonormal residuals of the alternate model,

$$WSS_1 = \sum_{i=p+q+1}^{n} \varepsilon_i^2, \ H_1. \tag{6.72}$$

We expect that

$$WSS_0 - WSS_1 \geq 0,$$

but the question is whether the difference is large enough to justify adopting of the model with the additional adjustable parameters.

We will consider the sampling variability of the relative improvement,

$$v = \frac{\frac{WSS_0 - WSS_1}{q}}{\frac{WSS_1}{n-p-q}}, \tag{6.73}$$

under the hypothesis that the simple model is the true one.

The expression of Equation (6.73) follows a well-studied distribution known as F, with q and $n - p - q$ parameters (called "degrees of freedom"), or $F(v; q, n - p - q)$. Thus, with a 5% probability of incorrectly rejecting the simple model, we judge that:

> The improvement in the fit through use of the more complex model is significant if
>
> $$v_{\text{exp}} > v(q, n - p - q; 0.95), \tag{6.74}$$
>
> where v_{exp} is the experimental value and $v(q, n - p - q; 0.95)$ is the 0.95 percentile of this distribution, which can be found in [109] and other references. Or, if
>
> $$WSS_0 - WSS_1 > \frac{q}{n-p-q} WSS_1 \, v(q, n - p - q; 0.95), \tag{6.75}$$
>
> then using the more complex model is warranted.

This is a rational answer to the question: How large must the improvement in the fit be in order to support the use of a model with additional parameters?

For small data sets, having more parameters to adjust allows us to reproduce the data much more faithfully, but this improvement in the fit does not necessarily promise better predictions. The variance ratio test takes the data size into account and specifies that the improvement in the fit has to be large to justify the rejection of the simple model in favor of a more complicated one.

Example 6.9 *$p = 1$, $q = 1$, $n = 50$, and we computed from the analysis of data $v_{\text{exp}} = 2.2$. From the table in [109] we find: $v(1, 48; 0.95) = 4.04$. Thus, we do not reject the simple model in favor of the more complicated model because the improvement from the simple model is not significant.*

Example 6.10 *Let us consider the data of Table 6.1. We use as basic model the intrinsic with power variogram (Equation (6.5)) and find $WSS_0 = 0.998$. Then, with the same variogram, we assume that the trend is linear,*

$$m(x) = \beta_1 + \beta_2 x, \tag{6.76}$$

and compute[1] that $WSS_1 = 4.4 \times 10^{-29}$. In this case, $n = 8$, $p = 1$, and $q = 1$. Then, the ratio is $v_{\text{exp}} = 1.4 \times 10^{29}$, indicating that the improvement from using the variable mean model is statistically significant. (This simply says that there is an astronomically low probability that 8 observations will fall exactly on a straight line if Equation (6.5) is valid.)

Exercise 6.1 *Consider some hydraulic head data, taken in horizontal flow in an aquifer and shown in Table 9.2. Fit a variogram for the intrinsic case. Then, check whether a linear-trend model gives a significantly better fit.*

6.15 Key points of Chapter 6

A variable can be represented as the sum of a deterministic part (drift or mean function) and a stochastic part. The drift is the sum of known functions with unknown coefficients. The best linear unbiased estimator is a linear function of the observations with weights such that the mean estimation error is zero (unbiasedness) and the mean square error is as small as possible. Restrictions imposed to eliminate the unknown coefficients to satisfy the unbiasedness requirement make part of the covariance function (of the stochastic part) redundant in estimation. That is, the covariance can be written as the sum of an essential part, called a generalized covariance function, and a redundant part, which may be

[1] The value should actually be 0 but the computed value is affected by the numerical accuracy of the computer.

neglected for purposes of linear unbiased estimation. In other words, the generalized covariance function is a simplified version of the customary covariance function.

The larger the part of the spatial variability described through the deterministic part, the simpler the generalized covariance that is needed and the less sensitive are the results on the generalized covariance function that is used. Therefore, one should not shy away from using variable-mean models only out of concern for difficulties associated with generalized covariance functions. The question is whether the data and other information can support the rather strong assumption of describing with a specific deterministic function part of the variability of a quantity that is measured at only a few places. The variance-ratio test provides some guidance on this matter.

A preliminary estimate of the generalized covariance can be obtained through the familiar method of fitting an equation to the experimental variogram of the detrended data. This estimate may then be improved by using the method of the orthonormal residuals, which is applied in the same way as for the constant-mean case (see Chapter 4).

7
More linear estimation

We now examine some other best linear unbiased estimators, including estimation with given mean or drift, estimation of the drift coefficients, estimation of the continuous part of a regionalized variable with a nugget, and estimation of spatial averages or volumes. Many of the formulae appearing in this chapter are summarized in reference [84].

7.1 Overview

So far we have seen how to apply best linear unbiased estimation (BLUE) when we have n observations $z(\mathbf{x}_1), \ldots, z(\mathbf{x}_n)$ and we want to find the value of z at location $\mathbf{x}_0$, with $z(\mathbf{x})$ being modeled as an intrinsic process. The method is known as *ordinary kriging* or just *kriging*. We also studied the same problem when the mean is variable, consisting of the summation of terms, each of which having one unknown drift coefficient; the approach that resulted is known as *universal kriging*.

By slightly changing the criteria or the conditions, variants of these estimators can be obtained. Furthermore, one can solve other problems; for example, one can estimate the volume or the spatial average of $z(\mathbf{x})$ over a certain domain or one can estimate the slope of $z(\mathbf{x})$ in a certain direction at some point [110]. There is no end to the list of problems in which we can apply estimation techniques.

In all of these cases, we apply best linear unbiased estimation. Once we set up the problem, this methodology is mathematically and algorithmically straightforward. In other words, we always follow the same basic steps and we always end up with a system of linear equations that can be solved with relatively little effort with currently available computers.

The steps to set up a problem are:

1. We express the unknown estimate as a linear function of the data. The weights or coefficients are found from the following two conditions.
2. We require unbiasedness.
3. We require minimum mean square error of estimation.

We will see several examples in this chapter. However, the algebra associated with setting up BLUE problems may be of little interest to most readers. For this reason, in this section, we will present a quick overview of the problems that we solve in detail in the other sections. Readers may then skip sections that contain applications they are not interested in.

7.1.1 Kriging with known mean

We examine the same problem as ordinary or universal kriging but we consider that the mean is known, *i.e.*, specified separately from the data. Then, we find an estimator that is a weighted average of the data and the given mean. This estimator uses the covariance function (not the variogram or the generalized covariance function).

An important special case is *simple kriging*, which involves a stationary function with known mean.

7.1.2 Estimation of drift coefficients

Although we seldom are interested in estimating the mean of a stationary function or the drift coefficients (in the linear model of Chapter 6), this is a straightforward problem. A potential application is in finding coefficients that can be used in data detrending.

7.1.3 Estimation of continuous part

Estimating the continuous part is an interesting variant of any kriging approach that may be considered when there is a nugget term. The difference from ordinary kriging is due to a different interpretation of what we try to estimate.

We can view the variable $z(\mathbf{x})$ as the summation of a continuous function $z_c(\mathbf{x})$ and a pure nugget effect process $\eta(\mathbf{x})$. The objective in ordinary kriging (OK) is to estimate a value of z, whereas the objective in continuous-part kriging (CPK) is to estimate z_c. In some applications, the nugget effect part represents undesirable "noise" that needs to be filtered out since we are really interested in the estimation of the continuous part.

The key differences are:

1. The CPK estimate yields the same result as the OK estimate everywhere except at exactly the location where there is a measurement. At a location of a measurement, OK reproduces the measurement, whereas the CPK does not. The OK estimate as a function of the location, $\hat{z}(\mathbf{x})$, is discontinuous at the location of an observation, but the CPK estimate, $\hat{z}_c(\mathbf{x})$, is continuous.
2. The mean square error in CPK is smaller than the mean square error in OK by the nugget-effect variance everywhere except at observation points. (The reason is that z_c is less variable than z.) At an observation point, the mean square error for OK vanishes, whereas for CPK it equals the nugget-effect variance.

7.1.4 Spatial averaging

We are often interested in estimating the average concentration of a solute over a block or the volume of rainfall or infiltration over an area. In all of these cases, the estimate involves an integral (area, volume, etc.). For example, we might seek an estimate of the average,

$$z_A = \frac{1}{|A|} \int_A z(\mathbf{x})\, d\mathbf{x}, \tag{7.1}$$

from point observations, $z(\mathbf{x}_1), \ldots, z(\mathbf{x}_n)$.

We follow the usual approach in which we set up a linear estimator with coefficients that are determined from the requirements of unbiasedness and minimum variance.

7.2 Estimation with known mean

7.2.1 Known drift

We consider the familiar interpolation problem: Given n measurements of regionalized variable z at locations $\mathbf{x}_1, \ldots, \mathbf{x}_n$, estimate the value of z at point $\mathbf{x}_0$. For example, estimate the piezometric head at a selected location, given measurements of head at observation wells. We assume that $z(\mathbf{x})$ is a realization of a random field with known mean function $m(\mathbf{x})$ and covariance function $R(\mathbf{x}, \mathbf{x}')$.

Linear estimation means that we limit our search to estimators that are a linear function of the measurements (including a constant):

$$\hat{z}_0 = \kappa + \sum_{i=1}^{n} \lambda_i z(\mathbf{x}_i), \tag{7.2}$$

where κ and λ_i, $i = 1, \ldots, n$, are deterministic coefficients. The difference from ordinary or universal kriging is that we add a constant that should take into account that the mean is known. Intuitively, this term represents the estimate if no observations are available.

Unbiased stands for the specification that $\kappa, \lambda_1, \ldots, \lambda_n$ should be selected so that the estimation error, weighted over all possible solutions, be zero:

$$E[\hat{z}_0 - z(\mathbf{x}_0)] = 0 \tag{7.3}$$

or

$$\kappa + \sum_{i=1}^{n} \lambda_i m(\mathbf{x}_i) - m(\mathbf{x}_0) = 0. \tag{7.4}$$

Substituting the value of κ, we get the following form for the estimator:

$$\hat{z}_0 = m(\mathbf{x}_0) + \sum_{i=1}^{n} \lambda_i (z(\mathbf{x}_i) - m(\mathbf{x}_i)). \tag{7.5}$$

This estimator may be interpreted as follows:

> The best estimate of z at $\mathbf{x}_0$ is $m(\mathbf{x}_0)$, which is the estimate prior to the data plus a correction prompted by the data. The correction from measurement i is given by the deviation of the observation $z(\mathbf{x}_i)$ from its prior estimate $m(\mathbf{x}_i)$ times a weight or influence coefficient λ_i.

Finally, minimum variance stands for the specification that the average of the square error of estimation

$$E[(\hat{z}_0 - z(\mathbf{x}_0))^2] = E\left[\sum_{i=1}^{n} \lambda_i (z(\mathbf{x}_i) - m(\mathbf{x}_i)) - (z(\mathbf{x}_0) - m(\mathbf{x}_0))\right]^2 \tag{7.6}$$

be as small as possible.

Expanding and then taking the expected values inside the summations, we get

$$E[(\hat{z}_0 - z(\mathbf{x}_0))^2] = \sum_{i=1}^{n}\sum_{j=1}^{n} \lambda_i \lambda_j R(\mathbf{x}_i, \mathbf{x}_j) - 2\sum_{i=1}^{n} \lambda_i R(\mathbf{x}_i, \mathbf{x}_0) + R(\mathbf{x}_0, \mathbf{x}_0). \tag{7.7}$$

Thus, the λs may be calculated from the solution of the well-defined optimization problem:

$$\min_{\lambda} \left\{ \sum_{i=1}^{n}\sum_{j=1}^{n} \lambda_i \lambda_j R(\mathbf{x}_i, \mathbf{x}_j) - 2\sum_{i=1}^{n} \lambda_i R(\mathbf{x}_i, \mathbf{x}_0) + R(\mathbf{x}_0, \mathbf{x}_0) \right\}. \tag{7.8}$$

Taking the derivative of this expression with respect to λ_i for $i = 1, \ldots, n$, we obtain a system of n linear equations with n unknowns:

$$\sum_{j=1}^{n} R(\mathbf{x}_i, \mathbf{x}_j)\lambda_j = R(\mathbf{x}_i, \mathbf{x}_0), \quad i = 1, \ldots, n. \tag{7.9}$$

When we solve this system of equations, we obtain the coefficients that define the linear estimator of Equation 7.5. The solution is unique as long as there are no redundant measurements. The mean square error then is

$$E[(\hat{z}_0 - z(\mathbf{x}_0))^2] = -\sum_{i=1}^{n} \lambda_i R(\mathbf{x}_i, \mathbf{x}_0) + R(\mathbf{x}_0, \mathbf{x}_0). \tag{7.10}$$

Let us now look at a special case.

7.2.2 Simple kriging

If $z(\mathbf{x})$ is stationary, the mean is a known constant and the covariance function is a function of the distance. Then, the estimator is

$$\hat{z}_0 = m + \sum_{i=1}^{n} \lambda_i (z(\mathbf{x}_i) - m). \tag{7.11}$$

From the kriging system we have

$$\sum_{j=1}^{n} R(\mathbf{x}_i - \mathbf{x}_j)\lambda_j = R(\mathbf{x}_i - \mathbf{x}_0), \quad i = 1, \ldots, n. \tag{7.12}$$

The mean square error then is

$$E[(\hat{z}_0 - z(\mathbf{x}_0))^2] = -\sum_{i=1}^{n} \lambda_i R(\mathbf{x}_i - \mathbf{x}_0) + R(0). \tag{7.13}$$

Example 7.1 *When the covariance function is a pure nugget effect,*

$$R(\mathbf{x}_i - \mathbf{x}_j) = \begin{cases} \sigma^2, & \text{if } \mathbf{x}_i = \mathbf{x}_j \\ 0, & \text{if } \mathbf{x}_i \neq \mathbf{x}_j \end{cases}, \tag{7.14}$$

then, assuming that the location of the unknown is not the same with the location of any of the measurements, the kriging system becomes

$$\lambda_i \sigma^2 = 0, \quad i = 1, \ldots, n, \tag{7.15}$$

which means that all the coefficients vanish,

$$\lambda_i = 0, \quad i = 1, \ldots, n, \tag{7.16}$$

and the mean square estimation error is

$$E[(\hat{z}_0 - z(\mathbf{x}_0))^2] = \sigma^2. \tag{7.17}$$

Notice that the estimator in ordinary kriging had a mean square estimation error of $\sigma^2 + \sigma^2/n$. The variance reduction in simple kriging is due to the

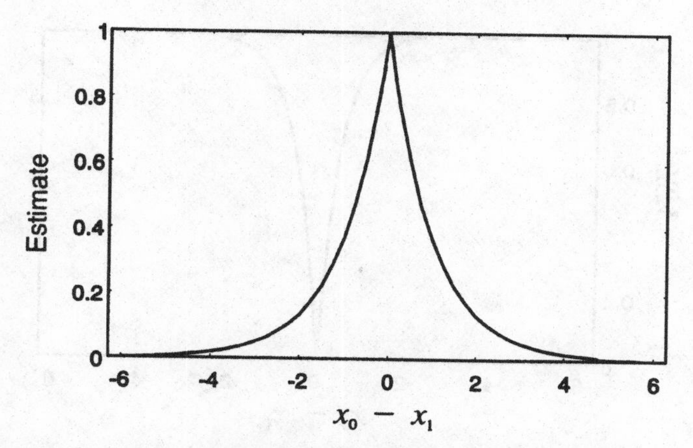

Figure 7.1 Simple kriging lambda coefficient as function of distance from observation.

fact that information about the mean, which is additional to the information contained in the data, is used in simple kriging.[1]

Example 7.2 *Consider an exponential covariance,*

$$R(\mathbf{x}_i - \mathbf{x}_j) = \sigma^2 \exp\left(-\frac{|\mathbf{x}_i - \mathbf{x}_j|}{l}\right), \tag{7.18}$$

with only one observation. Then,

$$\lambda_1 \sigma^2 = \sigma^2 \exp\left(-\frac{|\mathbf{x}_1 - \mathbf{x}_0|}{l}\right).$$

Thus the BLUE estimator is

$$\hat{z}_0 = m + \exp\left(-\frac{|\mathbf{x}_1 - \mathbf{x}_0|}{l}\right)(z(\mathbf{x}_1) - m), \tag{7.19}$$

with

$$MSE = \sigma^2\left(1 - \exp\left(-\frac{2|\mathbf{x}_1 - \mathbf{x}_0|}{l}\right)\right). \tag{7.20}$$

Figure 7.1 shows the λ coefficient and Figure 7.2 shows the mean square error of estimation as a function of $x_1 - x_0$ in a one-dimensional case, for $m = 0, \sigma^2 = 1$, and $l = 1$. One can see that the best estimate coincides with the observation at the location of the observation and with the mean far away (a distance of few ls) from the observation. The MSE is zero at the location where we have a measurement, and it tends to the variance away from the measurement.

[1] It is incorrect to use the data to estimate the mean and then to use simple kriging with the same observations because in simple kriging we assume that the mean represents information obtained separately from the n measurements.

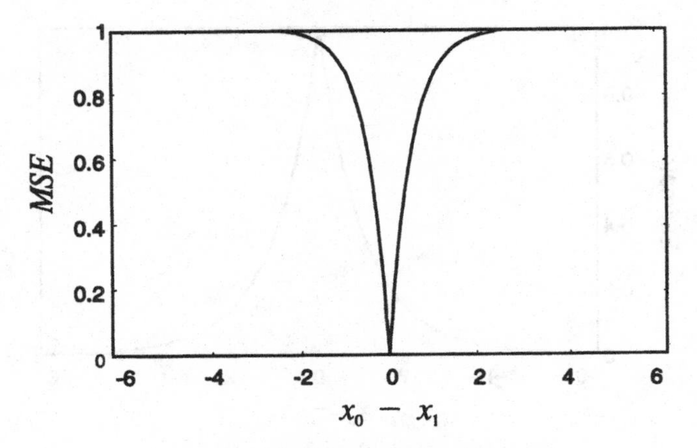

Figure 7.2 Mean square estimation error.

7.3 Estimation of drift coefficients

Let us use, for illustration purposes, the stationary case with unknown mean m and covariance function $R(h)$. We will use the n observations $z(\mathbf{x}_1), \ldots, z(\mathbf{x}_n)$ to infer the value of m.[2]

We will use a linear estimator,

$$\hat{m} = \sum_{i=1}^{n} \lambda_i z(\mathbf{x}_i). \tag{7.21}$$

Keep in mind that $\hat{m}$ is not the actual mean m but an estimate of it. The estimation error, $\hat{m} - m$, may be viewed as a random variable, because it depends on random variables $z(\mathbf{x}_1), \ldots, z(\mathbf{x}_n)$. For unbiasedness, we want the expected value to be zero:

$$E\left[\sum_{i=1}^{n} \lambda_i z(\mathbf{x}_i) - m\right] = 0 \tag{7.22}$$

or

$$\left(\sum_{i=1}^{n} \lambda_i - 1\right) m = 0. \tag{7.23}$$

For this condition to hold, we need to limit our attention to coefficients that satisfy

$$\sum_{i=1}^{n} \lambda_i = 1. \tag{7.24}$$

[2] The same general approach can be used to estimate the drift coefficients of the linear model used in Chapter 6.

The mean square error of estimation is

$$MSE(\hat{m}) = E\left[\left(\sum_{i=1}^{n} \lambda_i z(\mathbf{x}_k) - m\right)^2\right] = \sum_{i=1}^{n}\sum_{j=1}^{n} \lambda_i \lambda_j R(\mathbf{x}_i - \mathbf{x}_j). \quad (7.25)$$

Then, the coefficients are obtained by minimizing the mean square estimation error subject to the unbiasedness constraint. Using the method of Lagrange multipliers, the kriging system that determines the coefficients can be written as

$$\sum_{j=1}^{n} R(\mathbf{x}_i - \mathbf{x}_j)\lambda_j + \nu = 0, \quad i = 1, \ldots, n \quad (7.26)$$

$$\sum_{j=1}^{n} \lambda_j = 1. \quad (7.27)$$

Solving this system with $n + 1$ linear equations with $n + 1$ unknowns, we obtain the values of the $\lambda_1, \ldots, \lambda_n$ and ν. Then, we can compute the best estimate of the mean. The mean square error is

$$MSE(\hat{m}) = E\left[\left(\sum_{i=1}^{n} \lambda_i z(x_k) - m\right)^2\right] = -\nu. \quad (7.28)$$

Example 7.3 *Consider the pure nugget effect case (Equation (7.14)). Then, from the kriging system,*

$$\lambda_i \sigma^2 + \nu = 0, \quad i = 1, \ldots, n \quad (7.29)$$

$$\lambda_i = -\frac{\nu}{\sigma^2} = \frac{1}{n}, \quad i = 1, \ldots, n$$

$$\nu = -\frac{\sigma^2}{n} \quad (7.30)$$

$$MSE(\hat{m}) = \frac{\sigma^2}{n}. \quad (7.31)$$

Thus, in the case of pure nugget effect, the variance of estimation of the mean is inversely proportional to the total number of observations.

How accurate are the estimates of the mean? Consider a stationary function with covariance that has variance σ^2 with length parameter l. In this case, it will be useful to define the *effective number* of observations, n_e, as

$$n_e = \frac{\sigma^2}{MSE(\hat{m})}. \quad (7.32)$$

In a sense, the effective number of observations defines how many independent measurements are really available. It turns out that when the length parameter is about the same as the largest distance between observations, $L_{\max}$, then the effective number of measurements is nearly 1 no matter how many observations we really have. Increasing the number of observations does not really increase the effective number of observations (as long as the largest distance does not change).

Thus, it should be clear that when the variance σ^2 is large and the length parameter over largest separation distance, $l/L_{\max}$, is also large, it is impossible to obtain an accurate estimate of the mean. (However, this has little impact on the accuracy of interpolation.)

Finally, note that in this section we have assumed that the covariance function is known. What if the process is intrinsic with linear variogram (which is nonstationary)? Can we use the results of this section by substituting $-\gamma(h)$ where $R(h)$ appears, as we did for ordinary kriging? The answer is no; an intuitive interpretation is that the mean of a nonstationary process is not well defined and, consequently, it is not meaningful to try to estimate its mean square error.[3]

7.4 Continuous part

We have already seen that in ordinary kriging (OK) the estimate always reproduces the observations. This is true even when there is a nugget effect. However, if there is a nugget effect, the best estimate, $\hat{z}(\mathbf{x})$, from kriging is discontinuous at the location of an observation. Is this result reasonable?

The answer depends on what we want to estimate. If we truly want to calculate the value of $z(\mathbf{x})$, a discontinuous process, it makes sense to try to reproduce the measurements. However, the nugget effect usually represents measurement error and microvariability; one may then question whether reproducing the measurement serves any practical purpose. Instead, what is needed in many cases is to filter out measurement error and microvariability. This can be achieved by using the method described in this section.

7.4.1 Rationale

Consider a discontinuous intrinsic random field $z(x)$. That is, its variogram $\gamma(h)$ is discontinuous at the origin. It is instructive to view $z(x)$ as the sum of

[3] Nevertheless, the λ coefficients could be computed and used to obtain a crude sort of estimate of the mean, which is useful for data detrending.

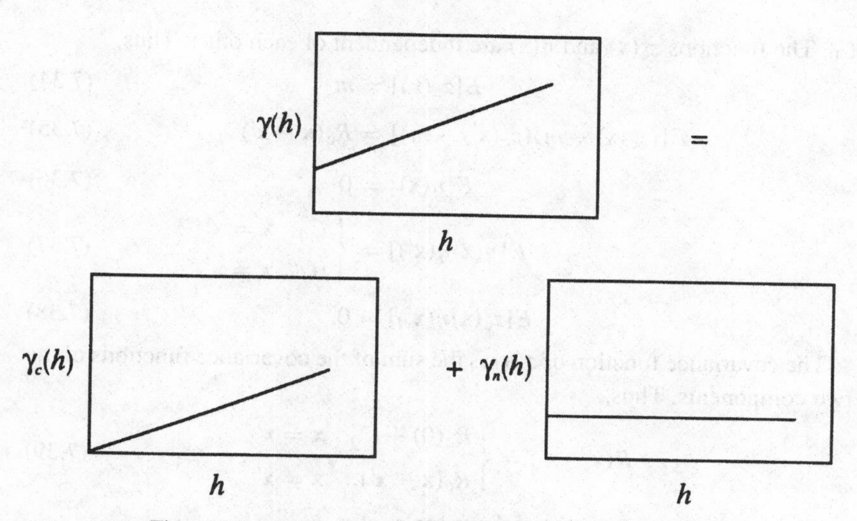

Figure 7.3 Variogram of z versus variograms of z_c and η.

two random fields, $z_c(\mathbf{x})$ and $\eta(\mathbf{x})$:

$$z(\mathbf{x}) = z_c(\mathbf{x}) + \eta(\mathbf{x}), \tag{7.33}$$

where $z_c(\mathbf{x})$ is a continuous intrinsic function and represents variability at the scales of typical distances between measurements and $\eta(\mathbf{x})$ is a discontinuous function with zero mean and no spatial correlation (pure nugget effect) and stands for random measurement error and spatial variability at a scale smaller than the distance between adjacent measurements. There is no correlation between z_c and η. Measurements are taken from a realization of the sum of the two processes, but we want to estimate the continuous function.

It is often appropriate to view the continuous part as the "signal" of interest; the measurements contain the signal plus "noise." The objective is to reproduce as accurately as possible the signal, not the corrupted-by-noise measurements.

Figure 7.3 shows the relation between the variogram of z and the variograms of z_c and η. The variogram of z_c passes through the origin, whereas the variogram of η is a pure nugget effect.

7.4.2 Kriging equations

It will be more convenient to work with covariance functions instead of variograms and, after the results are obtained, to replace R with $-\gamma$.

In Equation 7.33, $z_c(\mathbf{x})$ is the continuous part, a stationary function with continuous covariance function $R_c(\mathbf{x})$ (*i.e.*, $R_c(\mathbf{x} - \mathbf{x}') \to R(0)$ as $|\mathbf{x} - \mathbf{x}'| \to 0$), $\eta(\mathbf{x})$ is a zero-mean, pure-nugget-effect, stationary random field with variance

C_0. The functions $z_c(\mathbf{x})$ and $\eta(\mathbf{x})$ are independent of each other. Thus,

$$E[z_c(\mathbf{x})] = m \tag{7.34}$$

$$E[(z_c(\mathbf{x}) - m)(z_c(\mathbf{x}') - m)] = R_c(\mathbf{x} - \mathbf{x}') \tag{7.35}$$

$$E[\eta(\mathbf{x})] = 0 \tag{7.36}$$

$$E[\eta(\mathbf{x})\eta(\mathbf{x}')] = \begin{cases} C_0, & \mathbf{x} = \mathbf{x}' \\ 0, & \mathbf{x} \neq \mathbf{x}' \end{cases} \tag{7.37}$$

$$E[z_c(\mathbf{x})\eta(\mathbf{x}')] = 0. \tag{7.38}$$

The covariance function of $z(\mathbf{x})$ is the sum of the covariance functions of the two components. Thus,

$$R(\mathbf{x} - \mathbf{x}') = \begin{cases} R_c(0) + C_0, & \mathbf{x} = \mathbf{x}' \\ R_c(\mathbf{x} - \mathbf{x}'), & \mathbf{x} \neq \mathbf{x}' \end{cases}. \tag{7.39}$$

We seek a linear estimator of the value z_c at $\mathbf{x}_0$,

$$\hat{z}_{0c} = \sum_{i=1}^{n} \lambda_i z(\mathbf{x}_i), \tag{7.40}$$

such that, for any m,

$$E[\hat{z}_{0c} - z_c(\mathbf{x}_0)] = 0 \tag{7.41}$$

or

$$\sum_{i=1}^{n} \lambda_i = 1 \tag{7.42}$$

and the smallest possible variance of estimation,

$$E[(\hat{z}_{0c} - z_c(\mathbf{x}_0))^2] = \sum_{i=1}^{n}\sum_{j=1}^{n} \lambda_i \lambda_j R(\mathbf{x}_i - \mathbf{x}_j) - 2\sum_{i=1}^{n} \lambda_i R_c(\mathbf{x}_i - \mathbf{x}_0) + R_c(0). \tag{7.43}$$

Then, the kriging system of equations and the mean square error can be calculated from

$$\sum_{j=1}^{n} \lambda_j R(\mathbf{x}_i - \mathbf{x}_j) + \nu = R_c(\mathbf{x}_i - \mathbf{x}_0), \quad i = 1, 2, \ldots, n \tag{7.44}$$

$$\sum_{j=1}^{n} \lambda_j = 1 \tag{7.45}$$

$$E[0(\hat{z}_{0c} - z_c(x_0))^2] = -\nu - \sum_{i=1}^{n} \lambda_i R_c(\mathbf{x}_i - \mathbf{x}_0) + R_c(0). \tag{7.46}$$

Note that:

1. When $\mathbf{x}_i = \mathbf{x}_0$, the estimate does not coincide with measurement $z(\mathbf{x}_i)$. However, at points where there are no measurements, OK and CPK give the same estimate.
2. The mean square estimation error value computed from OK equations is larger than the value computed from CPK equations by C_0. This difference was to be expected since the former is the mean square error of estimation of $z(\mathbf{x}_0)$, whereas the latter is the mean square error of estimation of the less erratic $z_c(\mathbf{x}_0)$.

Comparison of the kriging equations indicates that:

- The left-hand side of the kriging system is the same for both versions, *i.e.*, whether z or z_c are sought.
- The difference is that in the right-hand side of the kriging equations and in the expression for the mean square error, R is used for estimation of z and R_c is used for estimation of z_c.

If the covariance function has infinite variance, the generalized covariance function or the variogram should be used. The expressions applicable in terms of variograms are:

$$-\sum_{j=1}^{n} \lambda_j \gamma(\mathbf{x}_i - \mathbf{x}_j) + \nu = -\gamma_c(\mathbf{x}_i - \mathbf{x}_0) - C_0, \quad i = 1, \ldots, n \quad (7.47)$$

$$\sum_{j=1}^{n} \lambda_j = 1 \quad (7.48)$$

$$E[(\hat{z}_{0c} - z_c(\mathbf{x}_0))^2] = -\nu + \sum_{i=1}^{n} \lambda_i \gamma_c(\mathbf{x}_i - \mathbf{x}_0). \quad (7.49)$$

***Example 7.4** Consider kriging with two measurements and assume that $\gamma(\mathbf{x} - \mathbf{x}') = C_0$, for $|x - x'| > 0$, while, as always, $\gamma(0) = 0$. In this case $\gamma_c = 0$ and we obtain $\lambda_1 = \lambda_2 = 1/2$ and $\nu = -C_0/2$ for any point. Thus, the estimate is the arithmetic average of the measurements, and the mean square estimation error is $C_0/2$.*

***Example 7.5** Consider the variogram*

$$\gamma(h) = \begin{cases} 0.03 + 0.75h, & h > 0 \\ 0, & h = 0 \end{cases}, \quad (7.50)$$

Table 7.1. *Data (observation and location)*

z	x
0.9257	0.10
−0.1170	0.30
0.0866	0.50
−0.5137	0.70
−0.3817	0.80

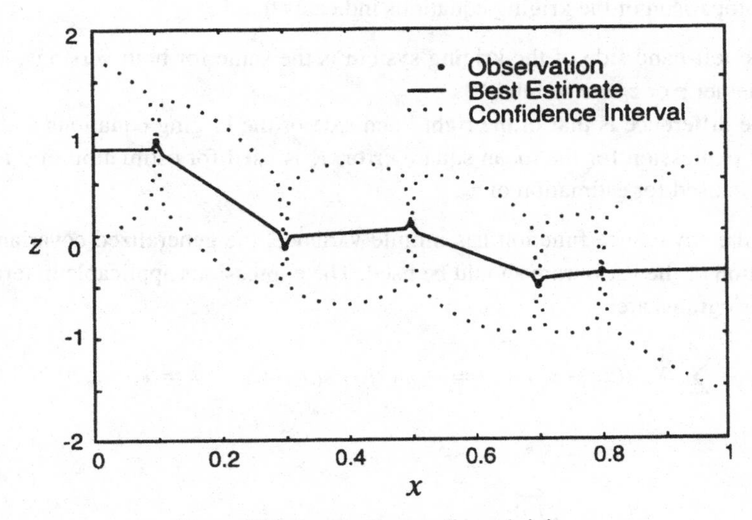

Figure 7.4 Results from ordinary kriging.

with observations given in Table 7.1. The results are given in Figures 7.4 for ordinary kriging and Figure 7.5 for continuous-part kriging. The best estimates (solid lines) are the same except at exactly the points where we have observations (indicated by solid circles). There, ordinary kriging reproduces the observations exactly, whereas continuous-part kriging reproduces the measurement only approximately. The reason for the difference is that CPK interprets the nugget effect as representing noise that must be filtered out of the observations. The confidence intervals (dotted lines) are also plotted. They are computed from $\hat{z} \pm 2\sqrt{MSE}$. *Note that, at an observation point, MSE is 0 for OK but is the value of the nugget variance for CPK. Everywhere else, the MSE for OK is larger than the MSE for CPK by the nugget variance.*

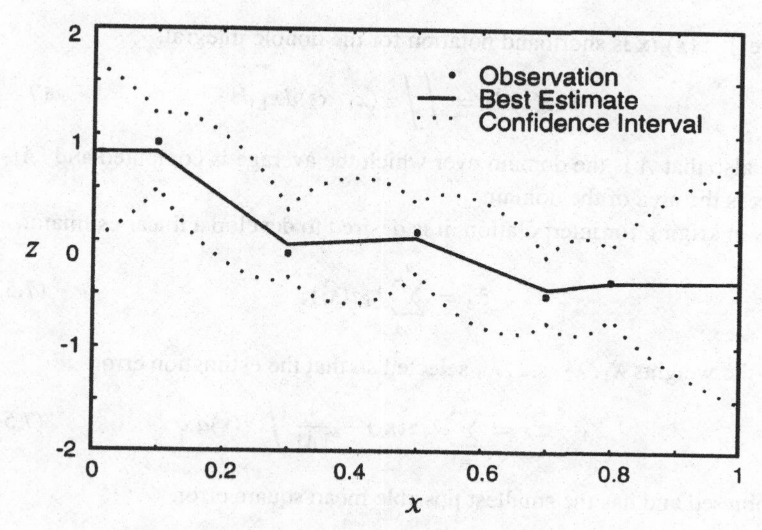

Figure 7.5 Results from continuous-part kriging.

7.5 Spatial averaging

Consider the problem of estimating spatial averages or volumes. Several examples come to mind:

- If $z(\mathbf{x})$ is infiltration rate, we are interested in calculating the average rate over a given area from point measurements obtained at locations $\mathbf{x}_1, \mathbf{x}_2, \ldots, \mathbf{x}_n$.
- If $z(\mathbf{x})$ represents concentration of a chemical, then we are interested in calculating the average concentration over a formation (the average concentration being needed in the estimation of total mass).
- One might want to estimate mean areal precipitation over a watershed from observations at rain gauges, which may be assumed to represent point measurements.

7.5.1 Stationary

To fix ideas, we will limit our discussion to regionalized variables defined on the plane and the problem of estimating the spatial average over a region, A, given measurements $z(\mathbf{x}_1), \ldots, z(\mathbf{x}_n)$. We will start by modeling $z(\mathbf{x})$ as a stationary function with constant (but unknown) mean m and given covariance function $R(\mathbf{x} - \mathbf{x}')$. However, the approach is more general.

The unknown quantity is

$$z_A = \frac{1}{|A|} \int_A z(\mathbf{x}) d\mathbf{x}, \tag{7.51}$$

where $\int_A z(\mathbf{x})d\mathbf{x}$ is shorthand notation for the double integral,

$$\int_A z(\mathbf{x})d\mathbf{x} = \int\int z(x_1, x_2)dx_1\, dx_2. \tag{7.52}$$

Note also that A is the domain over which the average is computed and $|A| = \int_A d\mathbf{x}$ is the area of the domain.

As in kriging for interpolation, it is desired to develop a linear estimator,

$$\hat{z}_A = \sum_{i=1}^{n} \lambda_i z(\mathbf{x}_i), \tag{7.53}$$

with the weights $\lambda_1, \lambda_2, \ldots, \lambda_n$ selected so that the estimation error

$$\hat{z}_A - z_A = \sum_{i=1}^{n} \lambda_i z(\mathbf{x}_i) - \frac{1}{|A|}\int_A z(\mathbf{x})d\mathbf{x} \tag{7.54}$$

is unbiased and has the smallest possible mean square error.

Unbiasedness It is required that

$$E[\hat{z}_A - z_A] = 0 \tag{7.55}$$

or

$$\sum_{i=1}^{n} \lambda_i\, E[z(\mathbf{x}_i)] - \frac{1}{|A|}\int_A E[z(\mathbf{x})]d\mathbf{x} = \left(\sum_{i=1}^{n} \lambda_i - 1\right) m = 0. \tag{7.56}$$

For this condition to hold for any value of m, it is required that

$$\sum_{i=1}^{n} \lambda_i = 1. \tag{7.57}$$

Minimum variance The variance of the estimation error is

$$\begin{aligned} E[(\hat{z}_A - z_A)^2] &= E\left[\left(\sum_{i=1}^{n} \lambda_i\, (z(\mathbf{x}_i) - m) - \frac{1}{|A|}\int_A (z(\mathbf{x}) - m)d\mathbf{x}\right)^2\right] \\ &= \sum_{i=1}^{n}\sum_{j=1}^{n} \lambda_i\lambda_j R(\mathbf{x}_i - \mathbf{x}_j) - 2\sum_{i=1}^{n} \lambda_i R_{Ai} + R_{AA}, \end{aligned} \tag{7.58}$$

where

$$\begin{aligned} R_{Ai} &= \frac{1}{|A|}\int_A R(\mathbf{x}_i - \mathbf{u})d\mathbf{u} \\ &= \frac{1}{|A|}\int\int R(x_{i1} - u_1, x_{i2} - u_2)du_1\, du_2 \end{aligned} \tag{7.59}$$

and

$$R_{AA} = \frac{1}{|A|^2}\int_A\int_A R(\mathbf{u}-\mathbf{v})d\mathbf{u}\,d\mathbf{v}$$
$$= \frac{1}{|A|^2}\iiiint R(u_1 - v_1, u_2 - v_2)du_1\,du_2\,dv_1\,dv_2. \quad (7.60)$$

A crude numerical method for the computation of these multiple integrals will be described in the next section. For the moment, assume that the integrals can be calculated using some analytical or numerical technique. Thus, R_{Ai} for $i = 1, \ldots, n$ and R_{AA} will now be considered known.

Minimizing (7.58) subject to the constraint (7.57), we obtain a system of $n+1$ linear equations:

$$\sum_{j=1}^{n} R(\mathbf{x}_i - \mathbf{x}_j)\lambda_j + \nu = R_{Ai}, \quad i = 1, \ldots, n$$
$$\sum_{j=1}^{n}\lambda_j = 1, \quad (7.61)$$

and the mean square estimation error is

$$E[(\hat{z}_A - z_A)^2] = -\nu - \sum_{i=1}^{n}\lambda_i R_{Ai} + R_{AA}. \quad (7.62)$$

7.5.2 Intrinsic

If the function is intrinsic rather than stationary, the same calculations hold if $R(\mathbf{h})$ is substituted by $-\gamma(\mathbf{h})$.[4]

The system of equations is

$$-\sum_{j=1}^{n}\gamma(\mathbf{x}_i - \mathbf{x}_j)\lambda_j + \nu = -\gamma_{Ai}, \quad i = 1, \ldots, n$$
$$\sum_{j=1}^{n}\lambda_j = 1, \quad (7.63)$$

and the mean square estimation error is

$$E[(\hat{z}_A - z_A)^2] = -\nu + \sum_{i=1}^{n}\lambda_i\gamma_{Ai} - \gamma_{AA}, \quad (7.64)$$

[4] This substitution works because the estimation error is an authorized increment, *i.e.*, it does not vary with the mean m.

where

$$\gamma_{Ai} = \frac{1}{|A|} \int_A \gamma(\mathbf{x}_i - \mathbf{u}) d\mathbf{u} \tag{7.65}$$

$$\gamma_{AA} = \frac{1}{|A|^2} \int_A \int_A \gamma(\mathbf{u} - \mathbf{v}) d\mathbf{u}\, d\mathbf{v}. \tag{7.66}$$

7.5.3 Variable mean

The approach can be extended rather easily to the general linear model that we saw in Chapter 6. In that model, the mean is variable but is modeled as the summation of p terms, each involving an unknown drift coefficient:

$$m(\mathbf{x}) = \sum_{k=1}^{p} f_k(\mathbf{x}) \beta_k, \tag{7.67}$$

where $f_k(\mathbf{x})$ are known functions and β_k are unknown drift coefficients, $k = 1, \ldots, p$.

Then, for each unknown drift coefficient we obtain an unbiasedness constraint,

$$\sum_{i=1}^{n} f_k(\mathbf{x}_i) \lambda_i = \frac{1}{|A|} \int_A f_k(\mathbf{u}) d\mathbf{u}, \quad k = 1, \ldots, p. \tag{7.68}$$

We also introduce p Lagrange multipliers. Thus, our kriging system comprises $n + p$ equations with $n + p$ unknowns:

$$\sum_{j=1}^{n} R(\mathbf{x}_i - \mathbf{x}_j) \lambda_j + \sum_{k=1}^{p} f_k(\mathbf{x}_i) \nu_k = R_{Ai}, \quad i = 1, \ldots, n \tag{7.69}$$

and Equations (7.68). The mean square estimation error is

$$E[(\hat{z}_A - z_A)^2] = -\sum_{k=1}^{p} \frac{1}{|A|} \int_A f_k(\mathbf{u}) d\mathbf{u}\, \nu_k - \sum_{i=1}^{n} \lambda_i R_{Ai} + R_{AA}. \tag{7.70}$$

7.6 Spatial averaging implementation

In this section, we will describe a simple method for computing the integrals and look at some examples.

7.6.1 Nugget effect

We will start by examining how to handle a discontinuity at the origin, *i.e.*, a nugget effect, when computing the integrals. Assume that $z(\mathbf{x})$ is stationary with covariance $R(h)$ that is discontinuous at the origin:

$$R(\mathbf{h}) = \begin{cases} C_0 + R_c(0), & \text{if } h = 0 \\ R_c(\mathbf{h}), & \text{if } h > 0 \end{cases}, \tag{7.71}$$

where C_0 is the nugget and $R_c(\mathbf{h})$ is a continuous covariance function (*i.e.*, $\lim_{h\to 0} R_c(h)$ tends to $R_c(0)$ as h tends to 0). For instance, if

$$R(\mathbf{h}) = \begin{cases} C_0 + \sigma^2, & \text{if } h = 0 \\ \sigma^2 \exp\left(-\frac{h}{l}\right), & \text{if } h > 0 \end{cases}, \tag{7.72}$$

then $R_c(\mathbf{h}) = \sigma^2 \exp(-\frac{h}{l})$.

Thus, $R(\mathbf{x}_i - \mathbf{u}) = R_c(\mathbf{x}_i - \mathbf{u})$ when $\mathbf{u}$ takes any value except $\mathbf{x}_i$ and $R(\mathbf{x}_i - \mathbf{u}) = C_0 + R_c(0)$ when $\mathbf{u}$ takes the value $\mathbf{x}_i$. Since the area associated with $\mathbf{x}_i$ is infinitesimal, the effect of C_0 is infinitesimal; consequently,

$$\int_A R(\mathbf{x}_i - \mathbf{u})d\mathbf{u} = \int_A R_c(\mathbf{x}_i - \mathbf{u})d\mathbf{u}. \tag{7.73}$$

Similarly,

$$\int_A \int_A R(\mathbf{u} - \mathbf{v})d\mathbf{u}\, d\mathbf{v} = \int_A \int_A R_c(\mathbf{u} - \mathbf{v})d\mathbf{u}\, d\mathbf{v}. \tag{7.74}$$

In other words, in computing the integrals R_{Ai} and R_{AA} needed in the estimation of the spatial averages, we may disregard the nugget effect and use only the continuous part of the covariance function.

Similarly, when we use the variogram

$$\gamma(\mathbf{h}) = \begin{cases} 0, & \text{if } \mathbf{h} = 0 \\ C_0 + \gamma_c(\mathbf{h}), & \text{if } \mathbf{h} \neq 0 \end{cases}, \tag{7.75}$$

then

$$\frac{1}{|A|}\int_A \gamma(\mathbf{x}_i - \mathbf{u})d\mathbf{u} = C_0 + \frac{1}{|A|}\int_A \gamma_c(\mathbf{x}_i - \mathbf{u})d\mathbf{u} \tag{7.76}$$

$$\frac{1}{|A|^2}\int_A \int_A \gamma(\mathbf{u} - \mathbf{v})d\mathbf{u}\, d\mathbf{v} = C_0 + \frac{1}{|A|^2}\int_A \int_A \gamma_c(\mathbf{u} - \mathbf{v})d\mathbf{u}\, d\mathbf{v}. \tag{7.77}$$

In numerical computations of the multiple integrals, it is essential to use R_c or γ_c.

7.6.2 Numerical quadrature

For illustration, let us consider the case that $z(\mathbf{x})$ is intrinsic with variogram $\gamma(\mathbf{h})$. The estimation of the kriging coefficients involves the calculation of n double integrals, $\int_A \gamma(\mathbf{x}_i - \mathbf{u})d\mathbf{u}$, and the calculation of the estimation error involves a quadruple integral, $\int_A \int_A \gamma(\mathbf{u} - \mathbf{v})d\mathbf{u}\, d\mathbf{v}$. Methods from calculus and numerical analysis may be used to calculate these deterministic integrals. In practice, one often has to resort to numerical methods.

A straightforward and general numerical method that may be adequate in many cases is to divide the total area into N subareas, $A_1, \ldots, A_N$. Each area A_k may be represented by a point $\mathbf{u}_k$ that may be the center of the area or some other representative point. Then,

$$|A| = \sum_{k=1}^{N} A_k \tag{7.78}$$

$$\int_A \gamma_c(\mathbf{x}_i - \mathbf{u})d\mathbf{u} = \sum_{k=1}^{N} \gamma_c(\mathbf{x}_i - \mathbf{u}_k)A_k \tag{7.79}$$

$$\int_A \int_A \gamma_c(\mathbf{u} - \mathbf{v})d\mathbf{u}\, d\mathbf{v} = \sum_{k=1}^{N} \sum_{l=1}^{N} \gamma_c(\mathbf{u}_k - \mathbf{u}_l)A_k A_l. \tag{7.80}$$

7.6.3 Examples

Let us study some applications. Consider the case of estimating mean areal precipitation from point measurements of rainfall.

Example 7.6 *Let us first look at the extreme case of uncorrelated measurements (pure nugget effect). In this case,*

$$\gamma(h) = \begin{cases} 0, & \text{if } h = 0 \\ C_0 & \text{if } h > 0 \end{cases}. \tag{7.81}$$

That is, $\gamma_c(h) = 0$. The system of Equation (7.64), becomes

$$\begin{array}{ccccccccc} 0 & -C_0\ \lambda_2 & -C_0\ \lambda_3 & \cdots & -C_0\ \lambda_n & + & \nu & = & -C_0 \\ -C_0\ \lambda_1 & 0 & -C_0\ \lambda_3 & \cdots & -C_0\ \lambda_n & + & \nu & = & -C_0 \\ -C_0\ \lambda_1 & -C_0\ \lambda_2 & 0 & \cdots & -C_0\ \lambda_n & + & \nu & = & -C_0 \\ & & \vdots & & & & & & \vdots \\ \lambda_1 & + \ \lambda_1 & + \ \lambda_3 & \cdots & + \ \lambda_n & & & = & 1. \end{array} \tag{7.82}$$

By symmetry, the solution is $\lambda_1 = \lambda_2 = \cdots = \lambda_n = 1/n$ and $\nu = -C_0/n$. The mean square error is given from Equation (7.64):

$$E[(\hat{z}_A - z_A)^2] = -\nu = \frac{C_0}{n}. \tag{7.83}$$

Thus, in the case of a pure nugget effect, the estimate is the arithmetic average of all measurements, and the mean square estimation error is inversely *proportional to the number of measurements. One should have anticipated this result since, according to the model, $z(\mathbf{x})$ varies at a scale much smaller than the distance between stations. Consequently, observations at all stations have the same weight in determining the areal mean. This is not the case if rainfall exhibits variability at a scale comparable to the distance between adjacent stations, as we will see in the next example.*

Example 7.7 *Assume that*

$$\gamma(h) = \begin{cases} 0, & \text{if } h = 0 \\ C_0 + h & \text{if } h \neq 0 \end{cases}. \tag{7.84}$$

The area (see Figure 7.6) is subdivided into 16 squares with sides equal to 2.5 km. Thus $|A_i| = 6.25\ km^2$. The coordinates of the center of each element are given in Table 7.2.

The location of the four stations is

$$\mathbf{x}_1 = \begin{bmatrix} 5 \\ 10 \end{bmatrix}, \quad \mathbf{x}_2 = \begin{bmatrix} 3.5 \\ 7.5 \end{bmatrix}, \quad \mathbf{x}_3 = \begin{bmatrix} 5 \\ 5 \end{bmatrix}, \quad \mathbf{x}_4 = \begin{bmatrix} 7.5 \\ 5 \end{bmatrix}$$

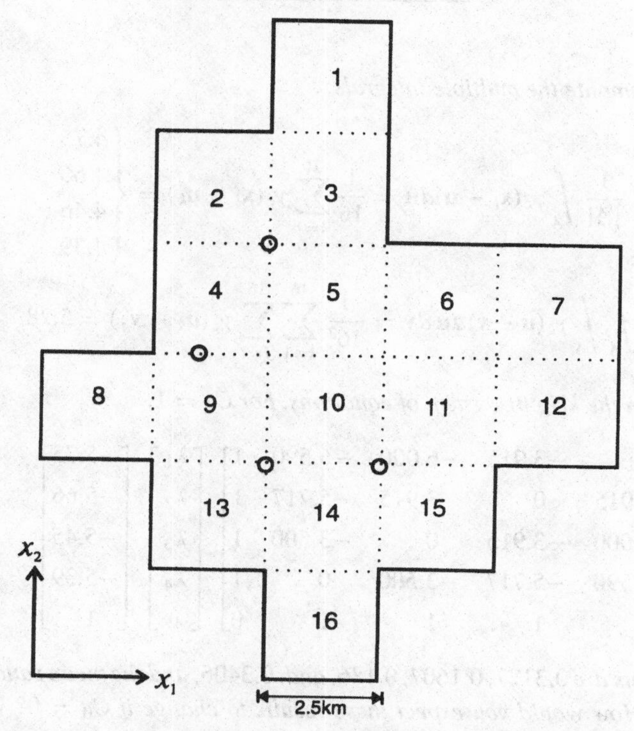

Figure 7.6 Map of area, showing subdivision into blocks and location of observations.

Table 7.2. *Coordinates of the center of each element*

i	u_1	u_2
1	6.25	13.75
2	3.75	11.25
3	6.25	11.25
4	3.75	8.75
5	6.25	8.75
6	8.75	8.75
7	11.25	8.75
8	1.25	6.25
9	3.75	6.25
10	6.25	6.25
11	8.75	6.25
12	11.25	6.25
13	3.75	3.75
14	6.25	3.75
15	8.75	3.75
16	6.25	1.25

First, compute the multiple integrals:

$$\frac{1}{|A|}\int_A \gamma_c(\mathbf{x}_i - \mathbf{u})d\mathbf{u} = \frac{1}{16}\sum_{k=1}^{16}\gamma_c(\mathbf{x}_i - \mathbf{u}_k) = \begin{cases} 4.75 \\ 4.66 \\ 4.46 \\ 4.39 \end{cases} \tag{7.85}$$

$$\frac{1}{|A|^2}\int_A\int_A \gamma_c(\mathbf{u} - \mathbf{v})d\mathbf{u}\,d\mathbf{v} = \frac{1}{16^2}\sum_{k=1}^{16}\sum_{j=1}^{16}\gamma_c(\mathbf{u}_k - \mathbf{v}_j) = 5.28. \tag{7.86}$$

Then, form the kriging system of equations. For $C_0 = 1$,

$$\begin{bmatrix} 0 & -3.915 & -6.000 & -6.590 & 1 \\ -3.915 & 0 & -3.915 & -5.717 & 1 \\ -6.000 & -3.915 & 0 & -3.500 & 1 \\ -6.590 & -5.717 & -3.500 & 0 & 1 \\ 1 & 1 & 1 & 1 & 0 \end{bmatrix} \begin{bmatrix} \lambda_1 \\ \lambda_2 \\ \lambda_3 \\ \lambda_4 \\ \nu \end{bmatrix} \begin{bmatrix} -5.75 \\ -5.66 \\ -5.45 \\ -5.39 \\ 1 \end{bmatrix}. \tag{7.87}$$

The weights are 0.3127, 0.1607, 0.186, *and,* 0.3406, *and the mean square error is* 1.04. *(How would you expect these results to change if* $C_0 = 0$? *Verify by repeating the calculations.)*

Now assume that it is proposed to move station 2 to either

$$\begin{bmatrix} 10 \\ 7.5 \end{bmatrix} \; or \; \begin{bmatrix} 7.5 \\ 7.5 \end{bmatrix}.$$

Which of the two positions would do the most to reduce the mean square error of estimation?

Because the accuracy depends only on the spatial structure of the function and the location of the gauges, the MSE of estimation can be calculated. We find that in the first location MSE $= 0.76$*, while in the second location MSE* $= 0.93$*. Consequently, the objective of accurate estimation of the mean areal rainfall is better met by moving the station to*

$$\begin{bmatrix} 10 \\ 7.5 \end{bmatrix}.$$

In this case, the weights are 0.31*,* 0.24*,* 0.26*, and* 0.19*.*

This example illustrates that the ability of these methods to evaluate the reliability of the estimates is very useful in the design of monitoring networks and of sampling strategies.

7.7 Key points of Chapter 7

Linear estimation methods are straightforward to develop and implement. In this chapter we got some practice applying them to several problems.

8
Multiple variables

This chapter introduces the subject of joint analysis of multiple spatial functions such as log-transmissivity and hydraulic head. Cokriging is the equivalent of kriging in the case of multiple variables. The general approach is introduced with a few comments on how to develop geostatistical models using only data. Additional information is given in references [1, 3, 7, 8, 32, 33, 35, 51, 59, 60, 64, 65, 87, 89, 100, 122, 123, 136, and 144].

8.1 Joint analysis

Some of the most interesting estimation problems involve two or more spatial functions. For example, in regional groundwater flow studies we deal with piezometric head (or pressure), transmissivity, and net recharge. Each of these quantities is variable in space. We have already seen how one may use log-transmissivity data to obtain a variogram or generalized covariance function and then linear minimum-variance unbiased estimates of log-transmissivity. The same procedure can be followed for the recharge and the head data. The relevant methods were described in Chapters 3 through 7.

However, what about the way each of these variables is correlated with the others? And how about using measurements of one type to estimate values of another type? For example, from groundwater mechanics we know that under certain conditions low values of transmissivity in a certain area tend to increase the slope of the piezometric head in the same area; an above-average elevation and curvature of the head surface correlates with increased rates of recharge in the same area; and so on. Such knowledge can be used to develop estimators of, say, log-transmissivity using measurements of not only log-transmissivity but also of head and recharge.

As we did previously, we will use probabilistic models because available information is not sufficient to specify a unique solution for the set of unknown spatial functions. Nevertheless, one can constrain these functions to belong to the families of realizations of correlated random fields. By choosing the first two moments of these random fields appropriately, one can represent some important characteristics of the unknown functions in the same way as for a single variable.

The procedure we will follow is a generalization of methods we became familiar with in previous chapters. Let us start by summarizing the basic steps:

1. The related functions are modeled as realizations of jointly distributed random fields.
2. The random fields are described through their joint first two moments. This "second-moment characterization" will require the introduction of the cross-covariance (and generalized cross-covariance) functions.
3. Following the second-moment characterization, prediction problems are solved using linear minimum-variance unbiased estimation.

8.2 Second-moment characterization

Consider the case of N correlated functions, $z_1(\mathbf{x}), \ldots, z_N(\mathbf{x})$, modeled as realizations of random fields. The first two moments of these fields comprise *mean functions*,

$$E[z_k(\mathbf{x})] = m_k(\mathbf{x}), \quad k = 1, \ldots, N, \tag{8.1}$$

autocovariance (or just *covariance*) *functions*,

$$E[(z_k(\mathbf{x}) - m_k(\mathbf{x}))(z_k(\mathbf{x}') - m_k(\mathbf{x}'))] = R_{kk}(\mathbf{x}, \mathbf{x}'), \quad k = 1, \ldots, N, \tag{8.2}$$

and *cross-covariance functions*,

$$E[(z_k(\mathbf{x}) - m_k(\mathbf{x}))(z_\ell(\mathbf{x}') - m_\ell(\mathbf{x}')] = R_{k\ell}(\mathbf{x}, \mathbf{x}'),$$
$$\text{for } k, \ell = 1, \ldots, N \text{ and } k \neq \ell. \tag{8.3}$$

Mean and autocovariance functions are familiar to us from Chapters 3 through 7 because they were also needed in the analysis of a single regionalized variable. The cross-covariance function is needed for the joint analysis of pairs of spatial functions.

Since the cross-covariance is the only new element in this description, let us study its properties and compare it with the autocovariance function. Unlike the autocovariance, the cross-covariance does not need to be symmetric in the arguments $\mathbf{x}$ and $\mathbf{x}'$. For example, the covariance between the head upgradient and the log-transmissivity downgradient may be completely different from the covariance between the log-transmissivity upgradient and the head downgradient. (The exact relation depends on the geometry of the domain and the boundary conditions.) However, note that

$$R_{k\ell}(\mathbf{x}, \mathbf{x}') = R_{\ell k}(\mathbf{x}', \mathbf{x}). \tag{8.4}$$

Furthermore, $R_{k\ell}(\mathbf{x}, \mathbf{x}), k \neq \ell$, is not the variance of a random variable but the covariance between two variables (say, head and recharge) at the same location in space. Consequently, unlike the autocovariance, it can have negative values. This is an important difference from the autocovariance.

Nevertheless, the autocovariance and cross-covariance functions that characterize a group of random fields must satisfy some conditions emanating from the requirement that variances cannot be negative. Consider, for example, the case of two jointly distributed random fields ($N = 2$). The first may be log-transmissivity and the second might be head. Let α be a linear combination of values from both fields, with n_1 values from variable z_1 and n_2 from variable z_2:

$$\alpha = \sum_{i=1}^{n_1} \lambda_{1i} z_1(\mathbf{x}_{1i}) + \sum_{i=1}^{n_2} \lambda_{2i} z_2(\mathbf{x}_{2i}). \tag{8.5}$$

The λs are prescribed coefficients. We can find the mean and variance of the random variable α from the mean, autocovariance, and cross-covariance functions of functions z_1 and z_2. The mean is

$$E[\alpha] = \sum_{i=1}^{n_1} \lambda_{1i} m_1(\mathbf{x}_{1i}) + \sum_{i=1}^{n_2} \lambda_{2i} m_2(\mathbf{x}_{2i}), \tag{8.6}$$

and the variance is

$$E[(\alpha - E(\alpha))^2] = \sum_{i=1}^{n_1} \sum_{j=1}^{n_1} \lambda_{1i} \lambda_{1j} R_{11}(\mathbf{x}_{1i}, \mathbf{x}_{1j}) + \sum_{i=1}^{n_2} \sum_{j=1}^{n_2} \lambda_{2i} \lambda_{2j} R_{22}(\mathbf{x}_{2i}, \mathbf{x}_{2j})$$

$$+ 2 \sum_{i=1}^{n_1} \sum_{j=1}^{n_2} \lambda_{1i} \lambda_{2j} R_{12}(\mathbf{x}_{1i}, \mathbf{x}_{2j}). \tag{8.7}$$

The point is that we must select R_{11}, R_{22}, and R_{12} so that for any n_1, n_2 and any λ weights, the expression of Equation (8.7) is nonnegative. If we do not, we may end up calculating a negative variance, which is absurd. Thus, as with autocovariance functions, not every model is acceptable.[1]

In addition to the requirement that the selected second-moment representation be mathematically reasonable, second moments must be selected so that they represent our understanding of the correlation structure of the spatial functions we are interested in. In principle, the determination of the second moments can be based on a purely statistical analysis of the data, which would hopefully suggest an appropriate model, or could make use of the relations between variables such as head and log-transmissivity as suggested by a physics-based mathematical model. We will address these issues after we look at the problem of best linear unbiased estimation for more than one spatial function.

[1] There are a number of ways to come up with covariance functions for a set of N correlated functions so that "positive definiteness" requirements are satisfied. One general way is to view each of the functions $z_1(\mathbf{x}), \ldots, z_N(\mathbf{x})$ as a linear transformation of N uncorrelated pure-nugget effect processes.

8.3 Best linear unbiased estimation

In this section, we will study the problem of linear estimation (also known as *cokriging*), assuming that we have the expressions for the first two moments. In other words, we will assume that the mean, autocovariance, and cross-covariance functions have somehow been determined. We will get back to the problem of how to develop a model and find its parameters in another section.

Up to this point we have seen best (minimum-variance) linear unbiased estimation applied only to a single variable. Actually, the generalization to the case of multiple variables is conceptually straightforward. To fix ideas while at the same time illustrating several concepts, let us illustrate the approach in the following example of joint analysis of head and log-transmissivity data. We assume we have n head measurements and m log-transmissivity data.

The model is as follows:

1. The mean F of the log-transmissivity Y is constant but unknown.
2. The mean $H(\mathbf{x})$ of the head ϕ is a known deterministic function of $\mathbf{x}$.
3. The autocovariance functions and the cross-covariance function of the two variables are known.

The problem is to use available measurements to estimate the value of log-transmissivity at a particular location $\mathbf{x}_0$.

Linear estimator Following the same general approach we have so many times applied to a single variable, we limit our attention to linear estimators. That is,

$$\hat{Y}_0 = \sum_{i=1}^{n} \mu_i \phi_i + \sum_{j=1}^{m} \lambda_j Y_j + \xi, \tag{8.8}$$

where ϕ is head and is Y is log-transmissivity. Note that now we have two sets of coefficients: those that multiply the head measurements and those that multiply the log-transmissivity measurements. The two sets of coefficients differ in a number of ways, *e.g.*, they have different units. We have also added a constant ξ, because the mean of the head is assumed known (the reason will become clearer later.)

The estimation problem is reduced to finding the coefficients μ, λ, and ξ. The estimation error, defined as the difference between the estimate and the actual value of Y_0 (or $Y(\mathbf{x}_0)$), is

$$\hat{Y}_0 - Y_0 = \sum_{i=1}^{n} \mu_i \phi_i + \sum_{j=1}^{m} \lambda_j Y_j + \xi - Y_0. \tag{8.9}$$

The coefficients will be determined according to the following specifications:

Unbiasedness The expected value of the estimation error must be zero or

$$E[\hat{Y}_0 - Y_0] = \sum_{i=1}^{n} \mu_i H_i + \left(\sum_{j=1}^{m} \lambda_j - 1 \right) F + \xi = 0. \tag{8.10}$$

For this condition to be satisfied for any value of F,

$$\sum_{j=1}^{n} \mu_j H_j + \xi \ = 0 \tag{8.11}$$

$$\sum_{j=1}^{m} \lambda_j - 1 \ = 0. \tag{8.12}$$

Note that (8.11) specifies the value of ξ and condition (8.12) eliminates the dependence of the estimate on the value of F. Thus, we may rewrite the estimator without ξ:

$$\hat{Y}_0 = \sum_{i=1}^{n} \mu_i (\phi_i - H_i) + \sum_{j=1}^{m} \lambda_j Y_j, \tag{8.13}$$

where the λ coefficients are constrained to satisfy Equation (8.12).

Minimum variance The coefficients must be such that the variance of the estimation error is as small as possible:

$$E[(\hat{Y}_0 - Y_0)^2] = E\left(\sum_{i=1}^{n} \mu_i (\phi_i - H_i) + \sum_{j=1}^{m} \lambda_j (Y_j - F) - (Y_0 - F) \right)^2, \tag{8.14}$$

where we made use of (8.12). Expanding and interchanging summation and expectation, we get

$$\begin{aligned} MSE &= E[(\hat{Y}_0 - Y_0)^2] \\ &= \sum_{i=1}^{n} \sum_{k=1}^{n} \mu_i \mu_k R_{\phi\phi}(\mathbf{x}_i, \mathbf{x}_k) + \sum_{j=1}^{m} \sum_{l=1}^{m} \lambda_j \lambda_\ell R_{YY}(\mathbf{x}_j, \mathbf{x}_\ell) \\ &\quad + R_{YY}(\mathbf{x}_0, \mathbf{x}_0) + 2 \sum_{i=1}^{n} \sum_{j=1}^{m} \mu_i \lambda_j R_{\phi Y}(\mathbf{x}_i, \mathbf{x}_j) \\ &\quad - 2 \sum_{i=1}^{n} \mu_i R_{\phi Y}(\mathbf{x}_i, \mathbf{x}_0) - 2 \sum_{j=1}^{m} \lambda_j R_{YY}(\mathbf{x}_j, \mathbf{x}_0). \end{aligned} \tag{8.15}$$

The coefficients λ and μ can thus be estimated from the solution of the following optimization problem: Minimize (8.15) subject to the constraint (8.12). Define the Lagrangian

$$L \equiv MSE + 2\nu \left(\sum_{j=1}^{m} \lambda_j - 1 \right). \tag{8.16}$$

Taking derivatives with respect to $\mu_i, i = 1, \ldots, n, \lambda_j, j = 1, \ldots, m$, and ν, we obtain the following system of $n + m + 1$ linear equations with $n + m + 1$ unknowns:

$$\sum_{k=1}^{n} \mu_k R_{\phi\phi}(\mathbf{x}_i, \mathbf{x}_k) + \sum_{j=1}^{m} \lambda_j R_{\phi Y}(\mathbf{x}_i, \mathbf{x}_j) = R_{\phi Y}(\mathbf{x}_i, \mathbf{x}_0), \quad i = 1, \ldots, n \tag{8.17}$$

$$\sum_{l=1}^{m} \lambda_\ell R_{YY}(\mathbf{x}_j, \mathbf{x}_\ell) + \sum_{i=1}^{n} \mu_i R_{\phi Y}(\mathbf{x}_i, \mathbf{x}_j) + \nu = R_{YY}(\mathbf{x}_j, \mathbf{x}_0), \quad j = 1, \ldots, m \tag{8.18}$$

$$\sum_{j=1}^{m} \lambda_j = 1. \tag{8.19}$$

Once this system is solved, the values of the coefficients can be substituted in the expressions for the best estimate and the mean square estimation error.

Exercise 8.1 *Consider the data and model described in Section 8.3. Derive the cokriging equations and the expression for the mean square error for the estimation of ϕ_0, the hydraulic head at location x_0. (Hint: Write the estimator*

$$\hat{\phi}_0 = \sum_{i=1}^{n} \mu_i \phi_i + \sum_{j=1}^{m} \lambda_j Y_j + \xi.$$

Next, show that as a consequence of the unbiasedness requirement, it can be written as

$$\hat{\phi}_0 = H_0 + \sum_{i=1}^{n} \mu_i (\phi_i - H_i) + \sum_{j=1}^{m} \lambda_j Y_j, \quad \textit{where} \quad \sum_{j=1}^{m} \lambda_j = 0.$$

Then write the expression for the MSE and apply the method of Lagrange multipliers.)

8.4 Another cokriging example

We will see here a variation of the cokriging example of Section 8.3.

We will repeat the procedure to find best estimates of the log-transmissivity as well as the head at any point under the assumption that the head mean is a linear function:

$$H(x_1, x_2) = \beta_1 + \beta_2 x_1 + \beta_3 x_2, \tag{8.20}$$

where x_1 and x_2 are spatial coordinates, and β_1, β_2, and β_3 are constant but unknown drift coefficients. All other conditions are identical with those given in the preceding section.

We will start with the estimation of log-transmissivity. The only difference from the equations in Section 8.3 is that $\xi = 0$, so that the following estimator is used:

$$\hat{Y}_0 = \sum_{i=1}^{n} \mu_i \phi_i + \sum_{j=1}^{m} \lambda_j Y_j. \tag{8.21}$$

The mean square estimation error expression is the same, but in the kriging system we have three more unbiasedness constraints and the corresponding Lagrange multipliers:

$$\sum_{k=1}^{n} \mu_k R_{\phi\phi}(\mathbf{x}_i, \mathbf{x}_k) + \sum_{j=1}^{m} \lambda_j R_{\phi Y}(\mathbf{x}_i, \mathbf{x}_j) + \nu_2 + \nu_3 x_{i1} + \nu_4 x_{i2} = R_{\phi Y}(\mathbf{x}_i, \mathbf{x}_0),$$
$$i = 1, \ldots, n \tag{8.22}$$

$$\sum_{l=1}^{m} \lambda_\ell R_{YY}(\mathbf{x}_j, \mathbf{x}_\ell) + \sum_{i=1}^{n} \mu_i R_{\phi Y}(\mathbf{x}_i, \mathbf{x}_j) + \nu_1 = R_{YY}(\mathbf{x}_j, \mathbf{x}_0),$$
$$j = 1, \ldots, m \tag{8.23}$$

$$\sum_{j=1}^{m} \lambda_j = 1 \tag{8.24}$$

$$\sum_{i=1}^{n} \mu_i = 0 \tag{8.25}$$

$$\sum_{i=1}^{n} \mu_i x_{i1} = 0 \tag{8.26}$$

$$\sum_{i=1}^{n} \mu_i x_{i2} = 0. \tag{8.27}$$

Next, consider the estimation of the hydraulic head. Following a similar approach, we write down a linear estimator,

$$\hat{\phi}_0 = \sum_{i=1}^{n} \mu_i \phi_i + \sum_{j=1}^{m} \lambda_j Y_j, \tag{8.28}$$

and impose unbiasedness and minimum variance requirements. The expression for mean square error of estimation is

$$MSE = \sum_{i=1}^{n} \sum_{k=1}^{n} \mu_i \mu_k R_{\phi\phi}(\mathbf{x}_i, \mathbf{x}_k) + \sum_{j=1}^{m} \sum_{l=1}^{m} \lambda_j \lambda_\ell R_{YY}(\mathbf{x}_j, \mathbf{x}_\ell)$$
$$+ R_{\phi\phi}(\mathbf{x}_0, \mathbf{x}_0) + 2 \sum_{i=1}^{n} \sum_{j=1}^{m} \mu_i \lambda_j R_{\phi Y}(\mathbf{x}_i, \mathbf{x}_j)$$
$$- 2 \sum_{i=1}^{n} \mu_i R_{\phi\phi}(\mathbf{x}_i, \mathbf{x}_0) - 2 \sum_{j=1}^{m} \lambda_j R_{Y\phi}(\mathbf{x}_j, \mathbf{x}_0). \tag{8.29}$$

The kriging system of equations is

$$\sum_{k=1}^{n} \mu_k R_{\phi\phi}(\mathbf{x}_i, \mathbf{x}_k) + \sum_{j=1}^{m} \lambda_j R_{\phi Y}(\mathbf{x}_i, \mathbf{x}_j) + \nu_2 + x_{1i}\nu_3 + x_{2i}\nu_4 = R_{\phi\phi}(\mathbf{x}_i, \mathbf{x}_0),$$
$$i = 1, \ldots, n \quad (8.30)$$

$$\sum_{l=1}^{m} \lambda_\ell R_{YY}(\mathbf{x}_j, \mathbf{x}_\ell) + \sum_{i=1}^{n} \mu_i R_{\phi Y}(\mathbf{x}_i, \mathbf{x}_j) + \nu_1 = R_{Y\phi}(\mathbf{x}_j, \mathbf{x}_0),$$
$$j = 1, \ldots m \quad (8.31)$$

$$\sum_{j=1}^{m} \lambda_j = 0 \quad (8.32)$$

$$\sum_{i=1}^{n} \mu_i = 1 \quad (8.33)$$

$$\sum_{i=1}^{n} \mu_i x_{i1} = x_{01} \quad (8.34)$$

$$\sum_{i=1}^{n} \mu_i x_{i2} = x_{02}. \quad (8.35)$$

Thus, we have seen that the problem of best linear unbiased estimation with many variables is conceptually and algorithmically straightforward.

We need to make a final remark before we leave the question of cokriging. Because the estimation error is an authorized increment, in the sense that it does not depend on unknown coefficients in the mean, we may replace the autocovariance or cross-covariance functions with their generalized equivalents (*i.e.*, with the parts of the covariance functions that do matter in cokriging).

8.5 Model selection

A mathematical model for the first two moments can be developed from data using an adaptation and extension of the approach used in Chapter 4. We will present a general outline.

Assume that we want to develop a model for log-transmissivity and head. First, let us deal with log-transmissivity only. On the basis of what is known and after inspection of the data, we may decide that an intrinsic model is appropriate. Using the methods of Chapter 4, we may determine the variogram and validate the model. With the hydraulic head, similarly, we may postulate a model, such as a first-order intrinsic function (*i.e.*, with linear drift). Using the methods of Chapter 6, we may estimate the parameters and validate the model using orthonormal residuals.

Then, the only task remaining is to determine a cross-covariance function. In the absence of better information, one may want to limit attention to cross-covariance functions that are a function of separation distance only, extending the concept of stationarity and isotropy. One can then plot the experimental cross-covariance. The experimental cross-covariance is conceptually similar to our familiar experimental variogram and is obtained following an analogous approach. In other words, form all pairs of head and log-transmissivity data. If there are n head measurements and m log-transmissivity measurements, we obtain $n \times m$ such pairs. For each pair we compute the product of the observations and the distance, and we then plot the product against the distance. Following a procedure similar to the one used to obtain the experimental variogram from the raw variogram, we obtain a line that is called the experimental cross-covariance.

However, graphical methods for selection of the cross-covariance are of limited help and cannot identify complex relations, so that one must rely mostly on the usual procedure of postulating a model and checking whether it is appropriate using validation techniques. In particular, the method of the orthonormal residuals is directly applicable. At the end of this chapter, we outline in compact matrix notation how to apply the equivalent of the method of covariance function parameter estimation through minimization of cR.

We will describe two methods that can lead to reasonable models in groundwater applications. The first method employs auxiliary variables and is outlined in Section 8.6. The second method, which is described in Chapter 9, makes use of mathematical groundwater flow and transport models.

8.6 Method of auxiliary variables

We will illustrate this method for developing a statistical model through an example from reference [68].

Assume that the objective is to come up with the contour map of the water table elevation of a shallow phreatic aquifer. The water table has been measured at a number of locations $z(x_1), \ldots, z(x_n)$. However, it is known that the water table is a subdued replica of the ground surface. Consequently, one can use measurements of the ground surface elevation to obtain details of the water table contour map that cannot be found in measurements of the piezometric head, available only in wells. This can be achieved by developing a model of the spatial correlation of the two functions and then applying cokriging.

The model was developed as follows: Based on data inspection, the water table function, $w(\mathbf{x})$, was represented as an intrinsic function with covariance function consisting of a nugget plus a linear term:

$$E[w(\mathbf{x}) - w(\mathbf{x}')] = 0 \tag{8.36}$$

$$K_{ww}(\mathbf{x} - \mathbf{x}') = \theta_{w1}\delta(\mathbf{x} - \mathbf{x}') + \theta_{w2}|\mathbf{x} - \mathbf{x}'|, \tag{8.37}$$

where

$$\delta(\mathbf{x} - \mathbf{x}') = \begin{cases} 1, & \text{if } \mathbf{x} = \mathbf{x}' \\ 0, & \text{otherwise} \end{cases};$$

$\theta_{w1} \geq 0$ and $\theta_{w2} \leq 0$ are parameters of the generalized covariance function K_{ww} (*i.e.*, a linear plus nugget variogram was adopted).

Similarly, for the ground surface elevation function $g(\mathbf{x})$,

$$E[g(\mathbf{x}) - g(\mathbf{x}')] = 0 \tag{8.38}$$

$$K_{gg}(\mathbf{x} - \mathbf{x}') = \theta_{g1}\delta(\mathbf{x} - \mathbf{x}') + \theta_{g2}|\mathbf{x} - \mathbf{x}'|, \tag{8.39}$$

where $\theta_{g1} \geq 0$ and $\theta_{g2} \leq 0$ are additional parameters.

We still have to determine the cross-covariance between the ground surface and the water table elevation. A simple approach [100] was used based on the introduction of an artificial spatial function t, defined as the sum of w and g:

$$t(\mathbf{x}) = w(\mathbf{x}) + g(\mathbf{x}). \tag{8.40}$$

This is represented as intrinsic

$$E[t(\mathbf{x}) - t(\mathbf{x}')] = 0 \tag{8.41}$$

$$K_{tt}(\mathbf{x} - \mathbf{x}') = \theta_{t1}\delta(\mathbf{x} - \mathbf{x}') + \theta_{t2}|\mathbf{x} - \mathbf{x}'|, \tag{8.42}$$

where $\theta_{t1} \geq 0$ and $\theta_{t2} \leq 0$ are two more parameters. These parameters can be estimated because the values of t are known at the locations of wells where both the ground surface and the water surface elevations are measured.

One can verify that

$$K_{tt}(\mathbf{x} - \mathbf{x}') = K_{ww}(\mathbf{x} - \mathbf{x}') + K_{gg}(\mathbf{x} - \mathbf{x}') + K_{wg}(\mathbf{x} - \mathbf{x}') + K_{wg}(-\mathbf{x} + \mathbf{x}'). \tag{8.43}$$

It is reasonable here[2] to take K_{gw} as an even (symmetric) function:

$$K_{wg}(\mathbf{x} - \mathbf{x}') = K_{wg}(-\mathbf{x} + \mathbf{x}'), \tag{8.44}$$

in which case,

$$K_{wg}(\mathbf{x} - \mathbf{x}') = \frac{1}{2}(K_{tt}(\mathbf{x} - \mathbf{x}') - K_{ww}(\mathbf{x} - \mathbf{x}') - K_{gg}(\mathbf{x} - \mathbf{x}')). \tag{8.45}$$

This example illustrates how the cross-covariance function can be found from the covariances of the original and auxiliary spatial functions. A similar result

[2] Generally, cross-covariances are not necessarily even functions. In other cases, they could actually be odd functions. Here, it makes sense to assume that the cross-covariance is even because there is no preferential direction of flow and the topography is rolling farmland.

would have been obtained if we had used as an auxiliary function the difference between the land surface elevation and the water table elevation. In practice, the auxiliary variables usually have some well-defined physical meaning.

Next, the parameters were estimated and the model was validated using residuals. Finally, the validated model was used for cokriging.

Such an approach is best suited for cases for which there are some clear a priori notions about what a model should look like and there is a reasonably good data base to verify the model. For cases such as the aforementioned application, it can be quite successful. However, for applications not meeting these requirements, such an approach may not be appropriate because it neglects a potentially important source of information: the principles that govern groundwater flow. We will address this issue in the next chapter.

8.7 Matrix form

We will complete this chapter with a synopsis of the best linear estimation equations in matrix form and a generalization of the method of parameter estimation through minimization of cR. This section is at a more advanced level and may be skipped at first reading.

8.7.1 Linear model

The observations are modeled as the sum of a deterministic and a stochastic term:

$$\mathbf{y} = \mathbf{X}\boldsymbol{\beta} + \boldsymbol{\epsilon}, \tag{8.46}$$

where $\mathbf{y}$ is the $n \times 1$ measurement vector, $\mathbf{X}$ is a known $n \times p$ matrix, $\boldsymbol{\beta}$ is a $p \times 1$ vector of parameters ("drift coefficients" in kriging), and $\boldsymbol{\epsilon}$ is a random vector with zero mean and covariance matrix $\mathbf{Q}_{yy}(\boldsymbol{\theta})$ (a function of some covariance parameters). Similarly, the unknowns are modeled as

$$\mathbf{y}_0 = \mathbf{X}_0\boldsymbol{\beta} + \boldsymbol{\epsilon}_0, \tag{8.47}$$

where $\mathbf{y}_0$ is the $m \times 1$ vector of unknowns, $\mathbf{X}_0$ is a known $m \times p$ matrix, and $\boldsymbol{\epsilon}_0$ is a random vector with zero mean, covariance matrix $\mathbf{Q}_{00}(\boldsymbol{\theta})$, and cross-covariance to $\boldsymbol{\varepsilon}$ that is $\mathbf{Q}_{0y}(\boldsymbol{\theta})$.

8.7.2 BLUE

The best estimate is

$$\hat{\mathbf{y}}_0 = \boldsymbol{\Lambda}\mathbf{y}, \tag{8.48}$$

where $\mathbf{\Lambda}$ is $m \times n$ and $\mathbf{M}$ is $p \times n$. They are found by solving

$$\begin{bmatrix} \mathbf{Q}_{yy} & \mathbf{X} \\ \mathbf{X}^T & \mathbf{0} \end{bmatrix} \begin{bmatrix} \mathbf{\Lambda}^T \\ \mathbf{M} \end{bmatrix} = \begin{bmatrix} \mathbf{Q}_{y0} \\ \mathbf{X}_0^T \end{bmatrix}, \tag{8.49}$$

where ^{T} indicates matrix transpose. The estimation error covariance matrix is

$$E[(\hat{\mathbf{y}}_0 - \mathbf{y}_0)(\hat{\mathbf{y}}_0 - \mathbf{y}_0)^T] = -\mathbf{X}_0\mathbf{M} + \mathbf{Q}_{00} - \mathbf{Q}_{0y}\mathbf{\Lambda}^T. \tag{8.50}$$

8.7.3 Parameter estimation

Select covariance parameters $\boldsymbol{\theta}$ that minimize

$$|\mathbf{Q}_{yy}|\ |\mathbf{X}^T\mathbf{Q}_{yy}^{-1}\mathbf{X}|\ |\mathbf{X}^T\mathbf{X}|^{-1}, \tag{8.51}$$

while at the same time satisfying

$$Q_2 = \frac{1}{n-p}\mathbf{y}^T\left(\mathbf{Q}_{yy}^{-1} - \mathbf{Q}_{yy}^{-1}\mathbf{X}\left(\mathbf{X}^T\mathbf{Q}_{yy}^{-1}\mathbf{X}\right)^{-1}\mathbf{X}^T\mathbf{Q}_{yy}^{-1}\right)\mathbf{y} = 1, \tag{8.52}$$

where $|\ |$ in this section stands for matrix determinant.

8.8 Key points of Chapter 8

This chapter dealt with estimation problems involving two or more correlated spatial functions. Our approach was to show how ideas familiar from previous chapters can be extended to the case of multiple functions.

The cross-covariance function represents the relation between different spatial functions. Given the covariance and cross-covariance functions, cokriging is a straightforward extension of the familiar BLUE approach. The estimator is a linear function of the measurements with weights that are obtained from the unbiasedness and minimum variance requirements. The weights are computed from a system of linear equations.

Model development in the case of multiple functions involves development of a model for the mean and covariance functions of each function and the cross-covariance for each function pair. As in earlier applications, we are guided by the data and what we know about the functions in question. Through example, we have illustrated that the cross-covariance function can sometimes be obtained from the covariance functions of the original and auxiliary functions. The method of orthonormal residuals can be used for parameter estimation and model validation.

9
Estimation and GW models

This chapter presents an introduction to how one can use estimation methods in conjunction with groundwater modeling. Applications include the calibration and validation of groundwater models and the evaluation of their predictive accuracy. A key objective is to illustrate the applicability of familiar principles to more challenging problems. The chapter also serves as an introduction to stochastic groundwater mechanics, which is covered in references such as [35, 53, 57, 58, 59, 60, 127, and 128].

9.1 Groundwater models

Groundwater (GW) models are mathematical representations of the flow of water and the transport of solutes in the subsurface, as in references [72 and 139]. Commonly, they employ finite-difference or finite-element approximations to the mathematical descriptions of hydrodynamic (flow, advection, dispersion) and physicochemical (*e.g.*, sorption, chemical transformations) processes.

Models are used to compute the hydraulic head, velocity, concentration, etc., from hydrologic and mass inputs, hydrogeologic and mass-transfer parameters, and conditions at the boundary of the domain. In the case of GW flow models, inputs include pumping, accretion, and evapotranspiration. The pertinent hydrogeologic parameters are hydraulic conductivity or transmissivity and storage coefficients. Boundary conditions are usually the head at the boundary of the flow domain or the flux rate through it. The output variable is hydraulic head (from which the discharge and advective velocity may also be calculated).

One of the difficulties encountered in applying these models to make predictions involves securing good estimates of the hydrogeologic parameters and, often, the boundary conditions. Most of the parameters are not measured directly. Hydrogeologic parameters are often estimated based on observations of how the system behaves under certain conditions. That is, parameters are selected so that they reproduce, when utilized in the numerical model, the measured values of hydraulic head, solute concentration, and other observable system variables. This approach has been widely applied as a manual trial and error procedure. In the past two decades, however, a number of systematic, computerized methods has been proposed [147]. In a sense, these methods treat the parameters as dependent variables in a new problem called the *inverse problem* of groundwater modeling. For example, see references [91, 103, 104, 133 and 138]. In practice, the problem of adjusting the parameters becomes an optimization problem in which parameter estimates are found from the minimization of a fitting criterion, as in references [17, 18, 25, and 26]. A good reference on general estimation methods is [124].

In numerical GW models, the domain is "discretized," *e.g.*, in finite-element models the geologic formation is conceptualized as a patchwork of uniform tiles or blocks known as elements. It is necessary to assign to every element parameter values. A numerical model may have hundreds or thousands of elements. If the formation is heterogeneous, the simple-minded approach is to let the parameters at each element of the domain be free to vary in any way that will allow the reproduction of the data. However, trying to estimate hundreds of unrestricted parameters from a few observations is asking for trouble. In a sense, groundwater models have too many parameters, so that [56, p. 337],

> ... in a given problem, many different sets of property estimates may provide satisfactory and essentially indistinguishable data fits. Some of these parameter estimates can be grossly in error with respect to the actual properties and as a result can lead to erroneous prediction of future reservoir behavior. To reduce the statistical uncertainty one must either decrease the number of unknowns or utilize additional information (about the parameters).

Overparameterization will likely cause an automatic calibration method to produce nonsensical results. For example, the estimates of the parameters may be different depending on the initial guess or the numerical accuracy of the computer. These undesirable features are accentuated by the fact that a small change in the measured hydraulic head may result in very different estimates of the conductivity parameters. For such problems, which are characterized as "ill-posed":

> A straightforward minimization without considering the uncertainty and reliability of the estimated parameters will result in meaningless parameter estimates.
>
> *[146, p. 665]*

The severity of the overparameterization problem has been recognized as the core of the parameter estimation problem, and several interesting approaches have been proposed ([74, 101, 125]). The most common method for reducing the number of independent parameters is the method of zonation, which is based on grouping the node parameters into zones. Another interesting approach is to represent the spatially distributed transmissivity as the sum of known basis functions with unknown coefficients. This, as we have seen, is the method used in geostatistics to represent large-scale variability. Some of these methods, such as that described in [146], address the need to choose the right number of parameters. However, trend surfaces and zones, although useful in representing large-scale variability, may be too inflexible to account for smaller-scale variability.

Emsellem and DeMarsily [49] argued that "conductivity is a parameter with no punctual value but with an average value in a region of a given size." They suggested that it makes sense to look for a solution that varies smoothly in some prescribed fashion.

The key issue is that, no matter what we do, the parameters cannot be determined uniquely. That is, even after taking stock of all available information, there remains an ensemble of equally plausible solutions. The conventional approach is to choose one of them and use it to predict the movement of a plume. The justification for choosing this parameter set is that it reproduces the data when it is used in the model. However, many other parameter sets have the same effect. When there are a hundred adjustable parameters, reproducing a dozen measurements is not a great accomplishment. Even physically implausible parameter sets may reproduce the data.

In this chapter, we will discuss an approach (from, [64, 67, 87, 122, 123]) that is an extension of the geostatistical methodologies that we have already seen. The problem of parameterization is addressed in the context of our familar structural analysis. To describe the spatial structure in a way that is not too restrictive and can account for variability at various scales, we employ geostatistical models. (Geostatistical methods in hydrogeology are also addressed in references [7, 35, 39, 40, and 41.]) These models include trend surfaces or zones, when appropriate, but also less restrictive representations based on the use of variograms and covariance functions. The number of independent parameters describing the aquifer properties can be reduced dramatically by employing geostatistical concepts. The geostatistical parameters of, for instance, transmissivity are estimated from point measurements of transmissivity (a common procedure in geostatistics) and point measurements of piezometric head, pumping rate, recharge, etc. The correlation between head and conductivity is computed using a groundwater model. The technique allows flexibility in the selection of the appropriate (given the data) geostatistical structure. After the geostatistical structure has been identified, the best linear unbiased estimate of the hydrogeologic parameter at any "point" of interest is found using best linear unbiased estimation theory. Here, we will focus on the basic approach but many extensions are possible, as in reference [27].

Exercise 9.1 *Discuss how the linear model of Equation (6.7) can accommodate the representation of zones and large-scale trends.*

9.2 Using a mathematical model

9.2.1 Equations and conditions

Consider, for example, the case of representing steady two-dimensional flow without sources or sinks, such as pumping wells or recharge. Application of conservation of mass and Darcy's law results in the following well-known partial differential equation:

$$\frac{\partial}{\partial x_1}\left(T\frac{\partial \phi}{\partial x_1}\right) + \frac{\partial}{\partial x_2}\left(T\frac{\partial \phi}{\partial x_2}\right) = 0, \tag{9.1}$$

where x_1 and x_2 are the spatial coordinates, T is transmissivity, a spatial function, and ϕ is the piezometric head, another spatial function. The location index

$$\mathbf{x} = \begin{bmatrix} x_1 \\ x_2 \end{bmatrix}$$

takes values in a two-dimensional domain.

There are also auxiliary conditions. For example, the numerical models that are used in practice require that the domain be finite in size and that on the boundary either the head or the flux be known (see Figure 9.1). This could be achieved by selecting the boundary to coincide with hydrologic features such as a river or lake in hydraulic connection with the groundwater system, where the head is approximately known, or a hydrologic divide where the flux is zero, etc.

In practice, we seldom have direct information to specify all the boundary conditions of a numerical groundwater model. The common approach is to

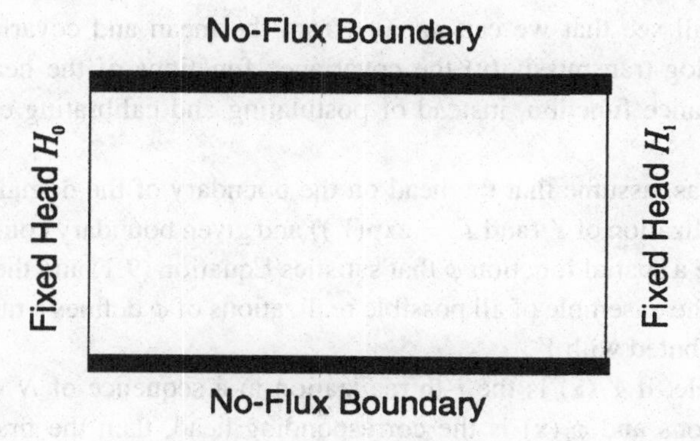

Figure 9.1 Example of a simple flow domain.

increase the size of the domain (to reduce the impact of the boundary conditions) and to calibrate the boundary conditions from other information, such as the level of wells in the interior of the domain. However, we will see that the approach followed here is much more attuned to the information that is really available.

Mathematical procedures are used to calculate $\phi(\mathbf{x})$ given $T(\mathbf{x})$ by solving the boundary value problem (*i.e.*, the partial differential equation, such as (9.1), and the prescribed boundary condition). Performing these calculations in an efficient and accurate manner is the subject of deterministic mathematical modeling.

9.2.2 Log-transmissivity

In geostatistical applications, it is more convenient to work with the logarithm of transmissivity, to be referred to as log-transmissivity:

$$Y = \ln(T). \tag{9.2}$$

The reasons have to do with assumptions implicit in linear estimation, which have been discussed in previous sections (see also [66]).

The governing equation can then be written in terms of Y:

$$\frac{\partial Y}{\partial x_1}\frac{\partial \phi}{\partial x_1} + \frac{\partial Y}{\partial x_2}\frac{\partial \phi}{\partial x_2} + \frac{\partial^2 \phi}{\partial x_1^2} + \frac{\partial^2 \phi}{\partial x_2^2} = 0. \tag{9.3}$$

We model $Y(\mathbf{x})$ as a random field with mean function $F(\mathbf{x})$ and covariance function $R_{YY}(\mathbf{x}, \mathbf{x}')$, as we saw in Chapter 3.

9.2.3 Derivation of joint moments

Here, we will see that we can *derive* (from the mean and covariance function of the log-transmissivity) the covariance functions of the head and the cross-covariance function, instead of postulating and calibrating expressions for them.

To fix ideas, assume that the head on the boundary of the domain is given. For each realization of Y (and $T = \exp(Y)$) and given boundary conditions one can compute a spatial function ϕ that satisfies Equation (9.1) and the boundary condition. The ensemble of all possible realizations of ϕ defines a random field jointly distributed with Y.

In principle, if $Y_i(\mathbf{x})$ is the i-th realization in a sequence of N equiprobable realizations and $\phi_i(\mathbf{x})$ is the corresponding head, then the first two moments of ϕ can be calculated by taking the limit of the following averages as

$N \to \infty$:

$$H(\mathbf{x}) = \frac{1}{N} \sum_{i=1}^{N} \phi_i(\mathbf{x}) \tag{9.4}$$

$$R_{\phi\phi}(\mathbf{x}, \mathbf{x}') = \frac{1}{N} \sum_{i=1}^{N} (\phi_i(\mathbf{x}) - H(\mathbf{x}))(\phi_i(\mathbf{x}') - H(\mathbf{x}')) \tag{9.5}$$

$$R_{\phi Y}(\mathbf{x}, \mathbf{x}') = \frac{1}{N} \sum_{i=1}^{N} (\phi_i(\mathbf{x}) - H(\mathbf{x}))(Y_i(\mathbf{x}') - F(\mathbf{x}')). \tag{9.6}$$

An advantage of this derivation is that we utilize principles of flow in porous media to derive mean and covariance functions instead of groping in the dark. In fact, even if some approximations have to be made to derive the moments of the dependent variable, this approach may be advantageous to the extent that it uses information that would be neglected otherwise. Notice that when one uses this approach, there is no need to make simplifications about the structure of the dependent variable, such as stationarity and isotropy.

In practice, the mean head, the covariance function of the head, and the head–log-transmissivity cross-covariance function are calculated using one of two methods:

- Monte Carlo simulations or
- first-order analysis.

These methods will be described next.

9.3 Monte Carlo simulations

9.3.1 The approach

The method of Monte Carlo (MC) simulations is conceptually simple and consists of commonly available tools:

1. A random-field generator. A random-field generator is a mathematical procedure that generates realizations of the input variable. For example, consider the log-transmissivity as a random function with zero mean and exponential covariance function, $\exp(-|\mathbf{x} - \mathbf{x}'|)$. A mathematical procedure can be developed to generate many possible functions $Y_1(\mathbf{x}), Y_2(\mathbf{x}), \ldots, Y_N(\mathbf{x})$. These functions are generated in such a way that $\frac{1}{N}\sum_{i=1}^{N} Y_i(\mathbf{x}) \simeq 0$ and $\frac{1}{N}\sum_{i=1}^{N} Y_i(\mathbf{x})Y_i(\mathbf{x}') \simeq \exp(-|\mathbf{x} - \mathbf{x}'|)$, provided that N is large.
2. A boundary-value problem solver. For each of the $Y_i(\mathbf{x})$ sample functions, one can solve a deterministic boundary value problem and obtain a head function $\phi_i(\mathbf{x})$.

3. Routines for the statistical analysis of data (Equations (9.4)–(9.6)). That is, one can take averages to approximate the mean, the autocovariance, and the cross-covariance.

The main advantage of the Monte Carlo method is its generality. In practice, because of the usually high cost of generating realizations of random fields and particularly of solving boundary value problems, N seldom exceeds a few hundred. The penalty for using a small N is that the results are affected by sampling error. (Sampling error in Monte Carlo simulations is due to the fact that the average value for a finite N differs from the average that would have been obtained at the limit of a very large N.) It can be shown that the standard (root mean square) sampling error, associated with estimation of the first two moments by taking averages over N realizations, is proportional to $1/\sqrt{N}$, so that beyond a certain point the standard sampling error decreases slowly. Nevertheless, even when the results are affected by sampling error, they can be adequate for purposes of best linear estimation (cokriging). An application can be found in reference [69].

9.3.2 Example

Consider one-dimensional flow between two boundaries where the head is known. Consider that the log-conductivity may be modeled as a realization of a stationary random field with mean m and exponential covariance function

$$R_{YY}(x - x') = \sigma^2 \exp(-|x - x'|/\ell), \tag{9.7}$$

where the variance σ^2 and the integral scale ℓ are known. The governing equation is

$$\frac{\partial Y}{\partial x}\frac{\partial \phi}{\partial x} + \frac{\partial^2 \phi}{\partial x^2} = 0, \tag{9.8}$$

with $\phi(0) = H_0$ and $\phi(L) = H_L$.

Analytical solutions to differential equations with variable coefficients are normally unavailable. Monte Carlo simulations usually require that the problem be discretized in some way, so that computer-based methods of solution can be used. Let us use a simple finite-difference scheme, with a uniform grid with spacing Δx, to approximate (9.8):

$$\frac{Y_{i+1} - Y_{i-1}}{2\Delta x}\frac{\phi_{i+1} - \phi_{i-1}}{2\Delta x} + \frac{\phi_{i+1} - 2\phi_i + \phi_{i-1}}{\Delta x^2} = 0, \quad \text{for } i = 2, \ldots, M-1 \tag{9.9}$$

or

$$[4 - (Y_{i+1} - Y_{i-1})]\phi_{i-1} - 8\phi_i + [4 + (Y_{i+1} - Y_{i-1})]\phi_{i+1} = 0 \tag{9.10}$$

and $\phi_1 = H_0$ and $\phi_M = H_L$.

The solution procedure is then the following:

1. Generate a sequence of M values $Y_1, Y_2, \ldots, Y_M$. For the exponential covariance function in one dimension, the following simple scheme can be used: Generate M standard normal variates $u_1, u_2, \ldots, u_M$. Then,

$$Y_1 = m + \sigma u_1 \tag{9.11}$$

$$Y_i = m + \rho(Y_{i-1} - m) + \sigma\sqrt{1-\rho^2}u_i, \quad \text{for } i = 2, \ldots, M, \tag{9.12}$$

where $\rho = \exp(-\Delta x/\ell)$. (This method is generally inapplicable for other covariance functions or for the exponential function in two or three dimensions.)
2. Solve the system of $M-2$ equations with $M-2$ unknowns, $\phi_2, \ldots, \phi_{M-1}$. In matrix notation, we write

$$\begin{bmatrix} -8 & 4+(Y_3-Y_1) & 0 & 0 & 0 & 0 \\ 4-(Y_4-Y_2) & -8 & 4+(Y_4-Y_2) & 0 & 0 & 0 \\ 0 & 4-(Y_5-Y_3) & -8 & 4+(Y_5-Y_3) & 0 & 0 \\ & & \vdots & & \vdots & \\ 0 & 0 & 0 & 0 & 4-(Y_M-Y_{M-2}) & -8 \end{bmatrix} \begin{bmatrix} \phi_2 \\ \phi_3 \\ \phi_4 \\ \vdots \\ \phi_{M-1} \end{bmatrix} = \begin{bmatrix} [-4+(Y_3-Y_1)]H_0 \\ 0 \\ 0 \\ \vdots \\ [-4-(Y_M-Y_{M-2})]H_L \end{bmatrix}. \tag{9.13}$$

(This approach is presented because it is general, but in this specific example one can take shortcuts based on the fact that the flow rate is constant everywhere.)

In the numerical results that will be presented, the normalized parameters are

$$H_0 = 1, \quad H_L = 0, \quad L = 1, \quad \sigma^2 = 1, \quad \ell = 0.5.$$

Because the head depends only on increments of the log-conductivity, we may use any value for the mean, such as $m = 0$.

Discretizing into $M = 21$ nodes and taking $N = 400$ realizations, we can determine the mean and covariance function of ϕ and the cross-covariance between ϕ and Y at the nodes. Some results are shown in Figures 9.2 through 9.6.

Figure 9.2 is the head mean, $H(x)$. It can be approximated closely by a straight line. Figure 9.3 shows the head variance, $R_{\phi\phi}(x, x)$, at each node. Clearly, $\phi(x)$ is not stationary since its variance is different from node to node. It is small

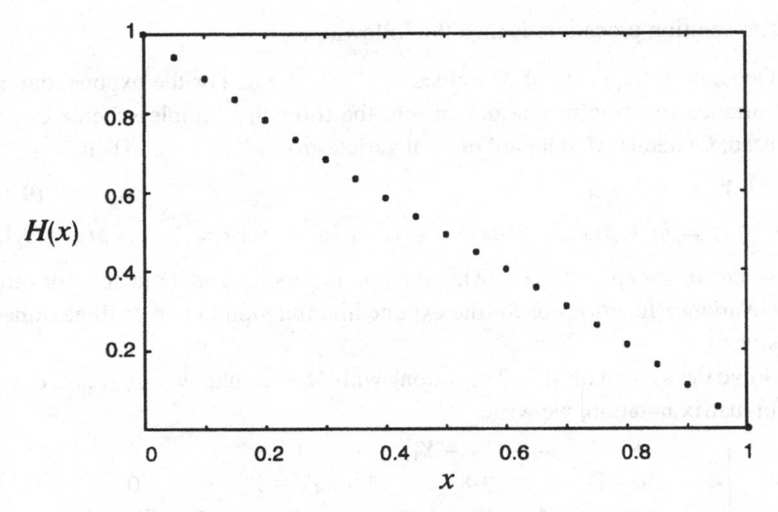

Figure 9.2 Mean head from Monte Carlo simulations.

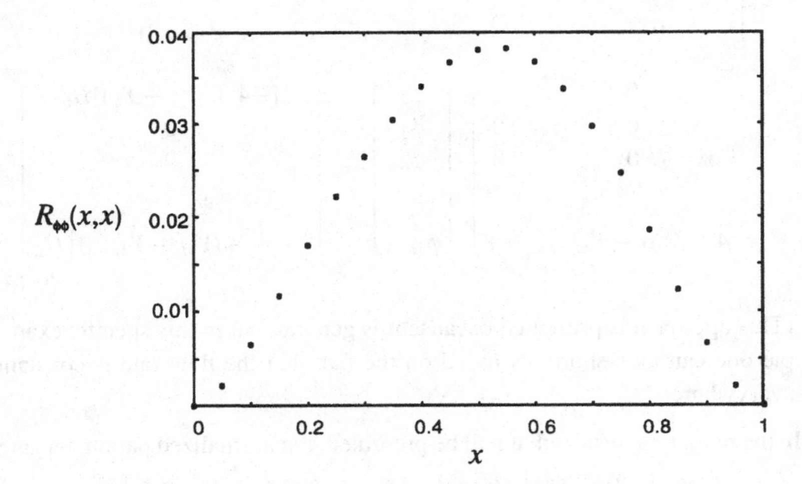

Figure 9.3 Variance of head from Monte Carlo simulations.

near the fixed-head boundaries and large at the center. Figure 9.4 depicts the correlation between the head at $x = 0.25$ and the head at any node. Figure 9.5 shows the cross-covariance of head with the log-conductivity at exactly the same location, $R_{\phi Y}(x, x)$.

Figure 9.6 is a plot of the correlation between the head at $x = 0.25$ and the log-K at any node. The result is as one should have expected: If the head at $x = 0.25$ is above the value given by the straight line of 9.2 at this location,

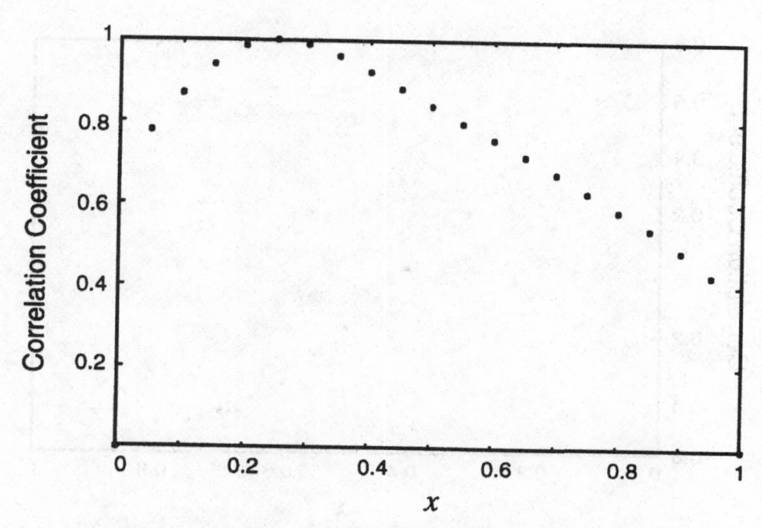

Figure 9.4 Correlation of head at x with head at $x = 0.25$.

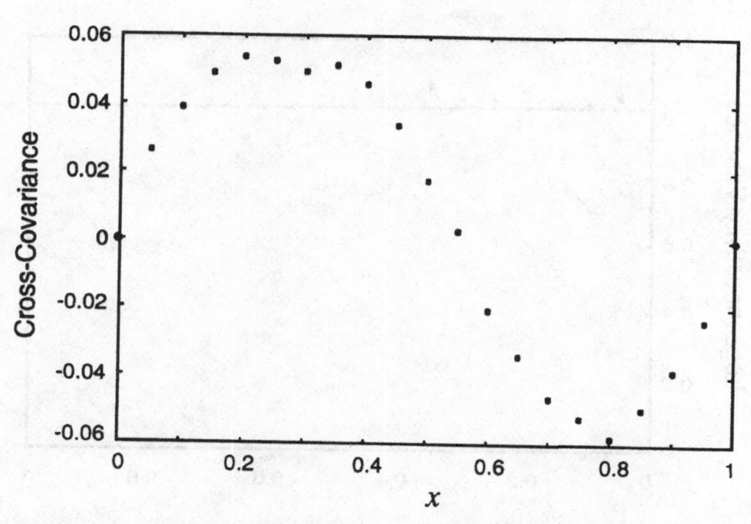

Figure 9.5 Cross-covariance of head–log-conductivity at same location.

then the conductivity near the upgradient end of the domain is likely to be high or the conductivity near the downgradient end of the domain is likely to be low.

Figure 9.7 shows $R_{YY}(x, x)$ computed from the 400 realizations. It can be seen that it is not exactly 1, as it would have been if N were infinitely large. This small discrepancy is due to the sampling error associated with MC simulations.

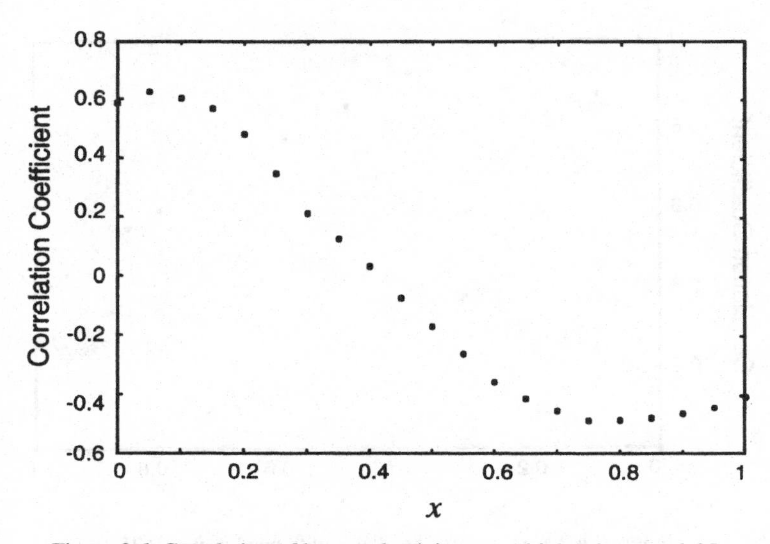

Figure 9.6 Correlation of log-conductivity at x with head at $x = 0.25$.

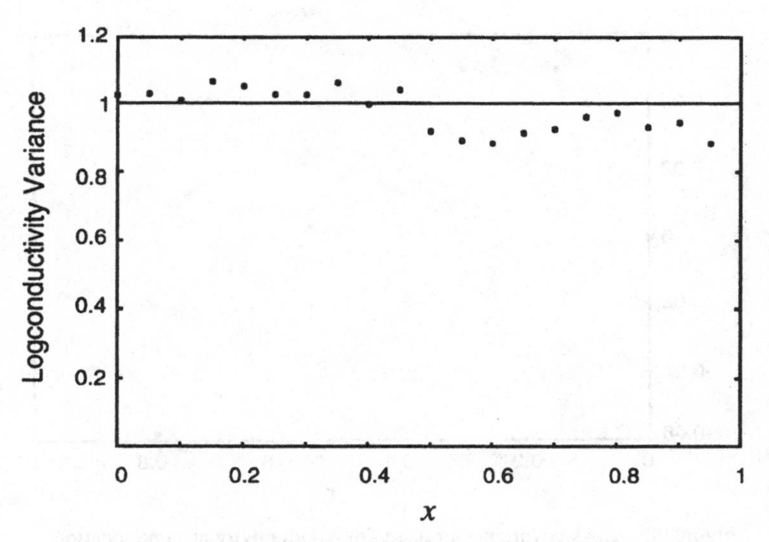

Figure 9.7 Log-conductivity variance from Monte Carlo simulations.

The results may vary substantially when the number of realizations is small, such as $N \sim 10$. In many estimation applications, the first two moments do not need to be calculated with great accuracy; $N = 400$ is sufficient in our example. Nevertheless, it is prudent to try to evaluate the size of the sampling error. Here are two methods:

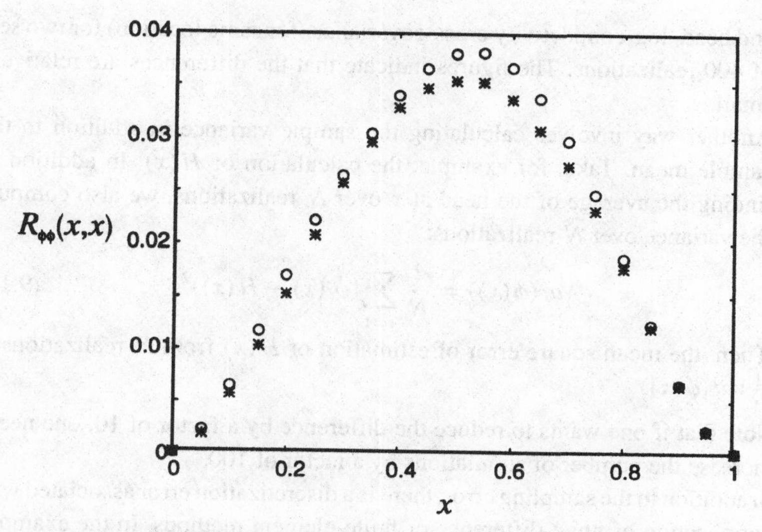

Figure 9.8 Comparison of head variance from two sets, each with 400 realizations.

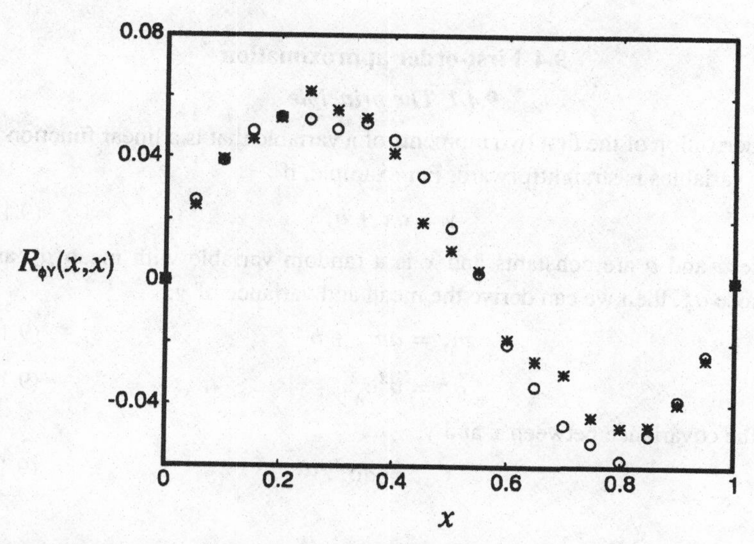

Figure 9.9 Comparison of head–log-conductivity covariance (same location) from two sets, each with 400 realizations.

1. One way is by repeating the procedure with a different set of realizations, *i.e.*, starting the simulation procedure with a different "seed number" (see Appendix C). The difference between the results provides an indication of the sampling error. For example, Figures 9.8 and 9.9 are the head variance

and head–log-conductivity cross-covariance (for same location) for two sets of 400 realizations. The figures indicate that the differences are relatively small.

2. Another way involves calculating the sample variance in addition to the sample mean. Take, for example, the calculation of $H(x)$. In addition to finding the average of the head at x over N realizations, we also compute the variance over N realizations:

$$Var(\phi(x)) = \frac{1}{N}\sum_{i=1}^{N}(\phi_i(x) - H(x))^2. \tag{9.14}$$

Then, the mean square error of estimation of $H(x)$ from N realizations is $\frac{1}{N}Var(\phi(x))$.

Note that if one wants to reduce the difference by a factor of 10, one needs to increase the number of simulations by a factor of 100.

In addition to the sampling error, there is a discretization error associated with the application of finite-difference or finite-element methods. In the example, Δx is one tenth of the integral scale l, which seems to be adequately small.

9.4 First-order approximation

9.4.1 The principle

The derivation of the first two moments of a variable that is a linear function of other variables is straightforward. For example, if

$$y = ax + b, \tag{9.15}$$

where a and b are constants and x is a random variable with mean μ_x and variance σ_x^2, then we can derive the mean and variance of y,

$$m_y = am_x + b \tag{9.16}$$

$$\sigma_y^2 = a^2\sigma_x^2, \tag{9.17}$$

and the covariance between x and y,

$$C_{xy} = a\sigma_x^2. \tag{9.18}$$

9.4.2 Linearization

We capitalize on this idea by developing an approximate method for calculating the first two moments of a dependent variable (*e.g.*, hydraulic head) from the first two moments of an independent variable (*e.g.*, log-transmissivity) after the relation between the two variables has been linearized. The approach is known as a *first-order analysis* or *small-perturbation approximation*.

Let

$$Y(\mathbf{x}) = F(\mathbf{x}) + f(\mathbf{x}) \tag{9.19}$$

$$\phi(\mathbf{x}) = H(\mathbf{x}) + h(\mathbf{x}), \tag{9.20}$$

where f and h are the fluctuations of the log-transmissivity and head about their mean functions F and H, respectively. We assume that the fluctuations of log-transmissivity and of head are small and, more precisely, are of the order of a small dimensionless scalar ε. Note that ε does not need to be interpreted as anything more than a guide in separating the "small" f and h terms from the "large" F and H terms and will play no other role in the analysis. In other words, all we need to know is that terms of order ε^2 are smaller than terms of order ε, which are smaller than terms of order 1, etc. That is,

$$f(\mathbf{x}) = O(\varepsilon) \tag{9.21}$$

$$h(\mathbf{x}) = O(\varepsilon), \tag{9.22}$$

where

$$\varepsilon \ll 1 \tag{9.23}$$

and O stands for "order of," which in an intuitive sense is about the same as "proportional to." Actually, this assumption is equivalent to linearizing the relation between f and h, since we say that they are both proportional to the same number. F and H are viewed as $O(1)$ (or $O(\varepsilon^0)$) terms. Substituting in the governing equation, we have

$$\frac{\partial(F+f)}{\partial x_1}\frac{\partial(H+h)}{\partial x_1} + \frac{\partial(F+f)}{\partial x_2}\frac{\partial(H+h)}{\partial x_2} + \frac{\partial^2(H+h)}{\partial x_1^2} + \frac{\partial^2(H+h)}{\partial x_2^2} = 0. \tag{9.24}$$

If we retain only zero-order terms ($O(\varepsilon^0)$), this reduces to

$$\frac{\partial F}{\partial x_1}\frac{\partial H}{\partial x_1} + \frac{\partial F}{\partial x_2}\frac{\partial H}{\partial x_2} + \frac{\partial^2 H}{\partial x_1^2} + \frac{\partial^2 H}{\partial x_2^2} = 0. \tag{9.25}$$

Equation (9.25) is a deterministic differential equation (since both F and H are deterministic functions) that can be solved subject to the given boundary conditions with the usual methods such as finite elements, finite diferences, etc.

Once H is calculated, we can equate terms of $O(\varepsilon^1)$ to get

$$\frac{\partial F}{\partial x_1}\frac{\partial h}{\partial x_1} + \frac{\partial F}{\partial x_2}\frac{\partial h}{\partial x_2} + \frac{\partial^2 h}{\partial x_1^2} + \frac{\partial^2 h}{\partial x_2^2} = -\frac{\partial f}{\partial x_1}\frac{\partial H}{\partial x_1} - \frac{\partial f}{\partial x_2}\frac{\partial H}{\partial x_2}. \tag{9.26}$$

If the value of the head on the boundary is given ($\phi = H$), the fluctuation is zero on the boundary.

Equation (9.26) is the *stochastic differential equation* that relates the fluctuation h to the fluctuation f. We call it stochastic because it relates stochastic processes: It can be used to determine the first two moments of h given the first two moments of f and appropriate boundary conditions. This is what we will mean by "solving a stochastic boundary value problem."

9.4.3 Moment derivation

The task is facilitated by the fact that Equation (9.26) denotes a linear relation between h and f. Also, if the boundary conditions to Equation (9.3) are deterministic, we have homogeneous boundary conditions,

$$h = 0 \quad \text{or} \quad \frac{\partial h}{\partial n} = 0 \text{ on the boundary,} \tag{9.27}$$

where n here is the direction perpendicular to the boundary. For a given function $f(\mathbf{x})$, the solution is of the form

$$h(\mathbf{x}) = \int G(\mathbf{x}, \mathbf{u}) f(\mathbf{u}) d\mathbf{u}, \tag{9.28}$$

where $G(\mathbf{x}, \mathbf{u})$ is the "influence" function. Note that here $\int$ indicates a multiple integral and that $d\mathbf{u}$ is shorthand notation for $du_1\, du_2 \ldots$.

Function $G(\mathbf{x}, \mathbf{u})$ quantifies the effect of fluctuation f at location $\mathbf{u}$ on fluctuation h at location $\mathbf{x}$. Its analytical derivation in a closed form may be anything but simple. Nevertheless, it is useful in understanding the principle to assume for a moment that $G(\mathbf{x}, \mathbf{u})$ has somehow been obtained. Then, the statistics of $h(\mathbf{x})$ can be calculated:

$$E[h(\mathbf{x})h(\mathbf{x}')] = E\left[\int\int G(\mathbf{x}, \mathbf{u}) G(\mathbf{x}', \mathbf{u}') f(\mathbf{u}) f(\mathbf{u}') d\mathbf{u}\, d\mathbf{u}'\right]. \tag{9.29}$$

Interchanging the order of expectation with integration, we obtain

$$E[h(\mathbf{x})h(\mathbf{x}')] = \int\int G(\mathbf{x}, \mathbf{u}) G(\mathbf{x}', \mathbf{u}') E[f(\mathbf{u}) f(\mathbf{u}')] d\mathbf{u}\, d\mathbf{u}' \tag{9.30}$$

$$= \int\int G(\mathbf{x}, \mathbf{u}) G(\mathbf{x}', \mathbf{u}') R_{YY}(\mathbf{u}, \mathbf{u}') d\mathbf{u}\, d\mathbf{u}'. \tag{9.31}$$

Similarly,

$$E[h(\mathbf{x}) f(\mathbf{x}')] = \int G(\mathbf{x}, \mathbf{u}) E[f(\mathbf{u}) f(\mathbf{x}')] d\mathbf{u} \tag{9.32}$$

$$= \int G(\mathbf{x}, \mathbf{u}) R_{YY}(\mathbf{u}, \mathbf{x}') d\mathbf{u}. \tag{9.33}$$

Thus, the first-order approximation method consists of four steps:

1. Linearization or small-perturbation approximation of the governing differential equation.
2. Solving a deterministic boundary value problem.
3. Obtaining influence function, $G(\mathbf{x}, \mathbf{u})$, which corresponds to the linear equation satisfied by the first-order terms.
4. Deriving the second moments of h by calculating integrals.

Useful closed-form analytical solutions can often be obtained by making a number of reasonable simplifications (examples are given in [8, 60, 32, 33, 35, and 57].) Commonly made assumptions include constant F and constant slope of H. It is also possible in simple cases to derive an analytical solution for a bounded domain with fixed-head or fixed-discharge boundary conditions (see [87]), but the analysis tends to be cumbersome. For this reason, most analytical solutions are derived for unbounded domains. For purposes of illustrating the methodology, this important special case will be described in the next section, after an example.

Example 9.1 *The applicability of the described methodology will be illustrated on the problem that we have already solved using the Monte Carlo method: steady flow in a one-dimensional aquifer with prescribed head boundary conditions and with no sources or sinks. In this case, the first-order equations lead to analytical closed-form solutions.*

The mean head function is

$$H(x) = H_0 - \Delta H \frac{x}{L}, \tag{9.34}$$

where $\Delta H = H_0 - H_L$, *and*

$$h(x) = -\frac{\Delta H}{L}\left[\frac{x}{L}\int_0^L f(u)du - \int_0^x f(u)du\right]. \tag{9.35}$$

We focus on the case that the covariance function of Y *is exponential:*

$$R_{ff}(x_i, x_j) = \sigma^2 \exp\left[-\frac{|x_i - x_j|}{l}\right]. \tag{9.36}$$

Then, after performing integrations that involve a fair amount of analytical work, it was derived [87] that $R_{\phi\phi}(x_i, x_j)$ *is given by the general relation,*

$$\begin{aligned} R_{\phi\phi}(x_i, x_j) &= E\{[\phi_i - H_i][\phi_j - H_j]\} \\ &= \Delta H^2 \sigma^2 \frac{l}{L} U\left(\frac{x_i}{L}, \frac{x_j}{L}, \frac{l}{L}\right), \end{aligned} \tag{9.37}$$

where

$$U\left(\frac{x_i}{L},\frac{x_j}{L},\frac{l}{L}\right)=2\,\frac{x_i}{L}\frac{x_j}{L}\left(1+\frac{l}{L}\exp\left(-\frac{L}{l}\right)-\frac{l}{L}\right)$$
$$-\frac{x_i}{L}\left[\frac{l}{L}\exp\left(-\frac{L}{l}\right)+\frac{l}{L}\exp\left(-\frac{x_j}{l}\right)\right.$$
$$\left.+2\,\frac{x_j}{L}-\frac{l}{L}-\frac{l}{L}\exp\left(-\frac{L-x_j}{l}\right)\right]$$
$$-\frac{x_j}{L}\left[\frac{l}{L}\exp\left(-\frac{L}{l}\right)+\frac{l}{L}\exp\left(-\frac{x_i}{l}\right)\right.$$
$$\left.+2\frac{x_i}{L}-\frac{l}{L}-\frac{l}{L}\exp\left(-\frac{L-x_i}{l}\right)\right]$$
$$+\left[\frac{l}{L}\exp\left(-\frac{x_i}{l}\right)+\frac{l}{L}\exp\left(-\frac{x_j}{l}\right)\right.$$
$$\left.+2\frac{\min(x_i,x_j)}{L}-\frac{l}{L}-\frac{l}{L}\exp\left(-\frac{|x_i-x_j|}{\ell}\right)\right]. \tag{9.38}$$

Also,

$$\begin{aligned}R_{\phi Y}(x_i,x_j)&=E\{[\phi_i-H_i][Y_j-F_j]\}\\&=\Delta H\,\sigma^2\frac{l}{L}V\left(\frac{x_i}{L},\frac{x_j}{L},\frac{l}{L}\right),\end{aligned} \tag{9.39}$$

where

$$V\left(\frac{x_i}{L},\frac{x_j}{L},\frac{l}{L}\right)=-\frac{x_i}{L}\left[2-\exp\left(-\frac{x_j}{l}\right)-\exp\left(-\frac{L-x_j}{l}\right)\right]$$
$$-\exp\left(-\frac{x_j}{l}\right)+\begin{cases}2-\exp\left(-\frac{|x_j-x_i|}{l}\right), & \text{if } x_i>x_j\\ \exp\left(-\frac{|x_j-x_i|}{l}\right), & \text{if } x_i<x_j\end{cases}. \tag{9.40}$$

Finally, $R_{\phi Y}(x_i,x_j)=R_{Y\phi}(x_j,x_i)$. The moments computed from the first-order analysis for this case are accurate for small σ^2. In practice, they are adequately accurate for purposes of cokriging even when σ^2 is large (such as 2 to 4).

A comparison betweeen Monte Carlo simulations (400 realizations) and first-order analysis is given in Figures 9.10–9.14. The asterisks are the results of the Monte Carlo simulations, the continuous lines are the results of

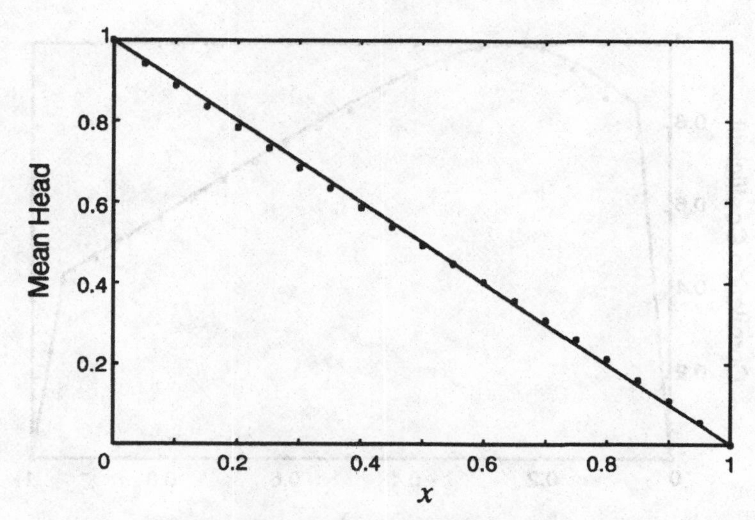

Figure 9.10 Comparison for mean head ($\sigma^2 = 1$).

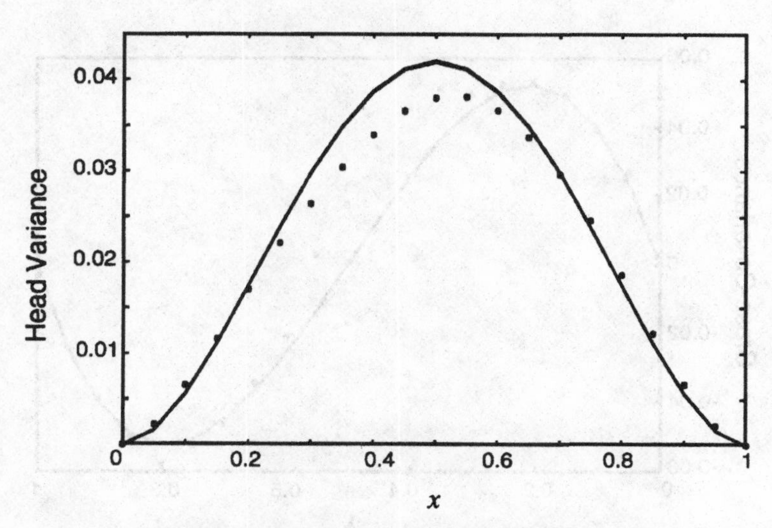

Figure 9.11 Comparison for head variance ($\sigma^2 = 1$).

the first-order analytical solution, and the dots (practically indistinguishable from the continuous lines) are the results of the numerical first-order analysis that we will describe later. Figure 9.10 compares the head mean, $H(x)$. Figure 9.11 shows the head variance, $R_{\phi\phi}(x, x)$, at each node. Figure 9.12 is the correlation between the head at $x = 0.25$ and the head at any node.

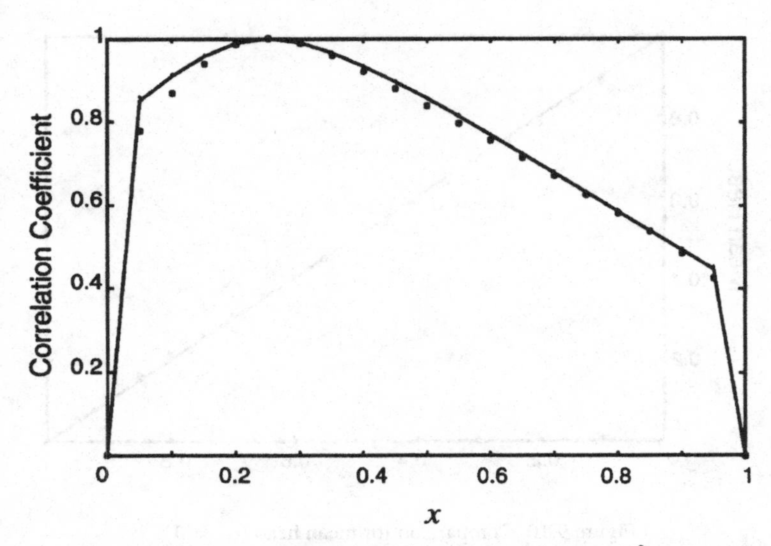

Figure 9.12 Comparison of $\phi(0.25)$ to $\phi(x)$ correlation ($\sigma^2 = 1$).

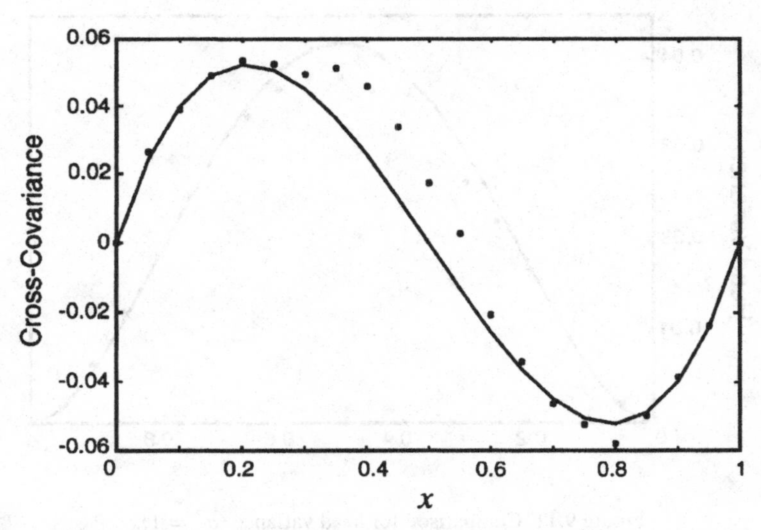

Figure 9.13 Comparison of cross-covariance ($\sigma^2 = 1$).

In Figure 9.13, the cross-covariance of head with the log-conductivity at exactly the same location, $R_{\phi Y}(x, x)$, is plotted. Figure 9.14 shows the correlation between the head at $x = 0.25$ and the log-conductivity at any node. From the comparison, we can see that the two methods produce very similar results.

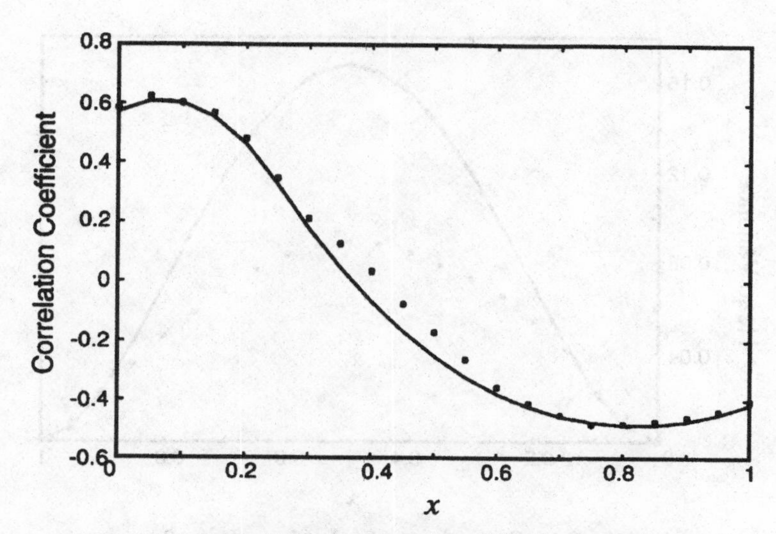

Figure 9.14 Comparison of ϕ (0.25) to $Y(x)$ correlation ($\sigma^2 = 1$).

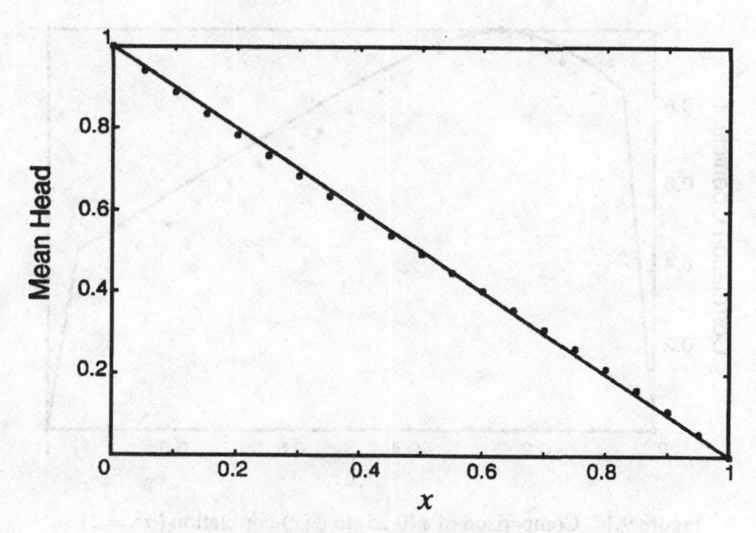

Figure 9.15 Comparison of mean head ($\sigma^2 = 2$).

The two methods were also compared for the case in which $\sigma^2 = 2$; these results are shown in Figures 9.15–9.19. It is evident that the differences increase with σ^2. For $\sigma^2 = 2$, the results of the perturbation method are less accurate but still adequate for many applications. Note that there is a general tendency for

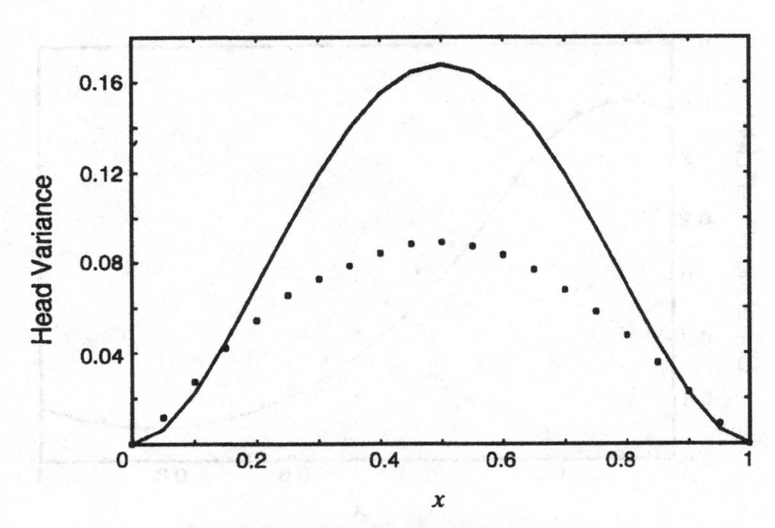

Figure 9.16 Comparison of head variance ($\sigma^2 = 2$).

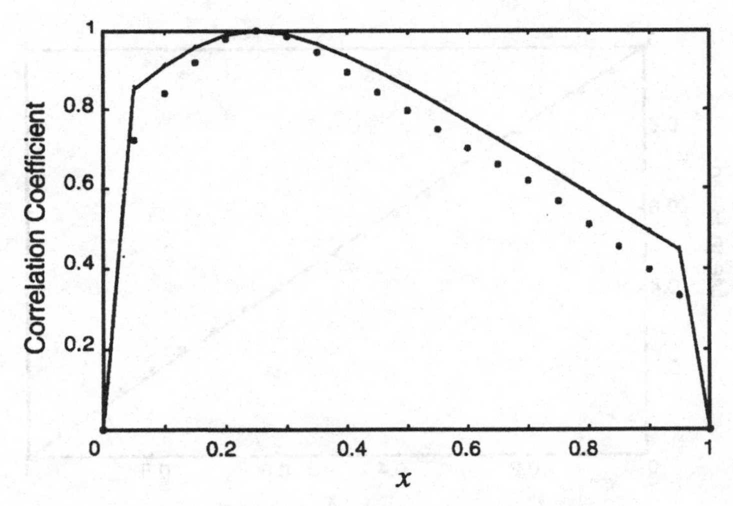

Figure 9.17 Comparison of $\phi(0.25)$ to $\phi(x)$ correlation ($\sigma^2 = 2$).

the head variance calculated from the perturbation method to be larger than the one computed from Monte Carlo simulations. (Keep in mind, though, that sampling error in the Monte Carlo method also tends to increase with the variance.) The agreement in the calculated cross-covariances and correlation coefficients is satisfactory.

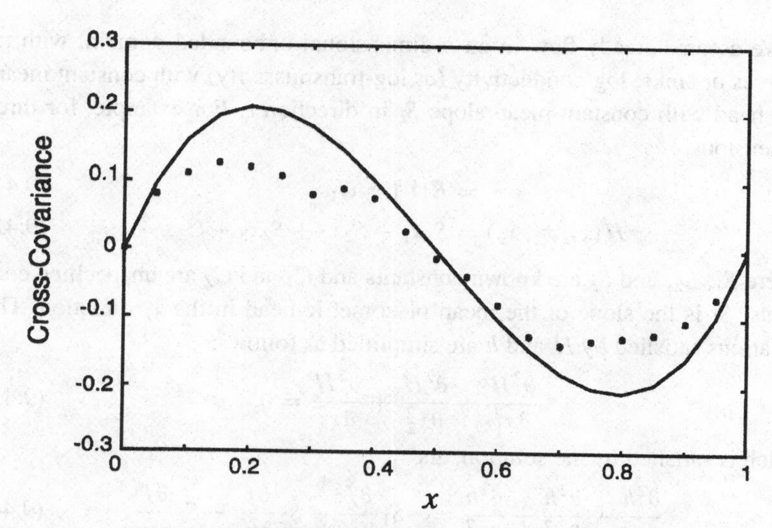

Figure 9.18 Comparison of cross-covariance ($\sigma^2 = 2$).

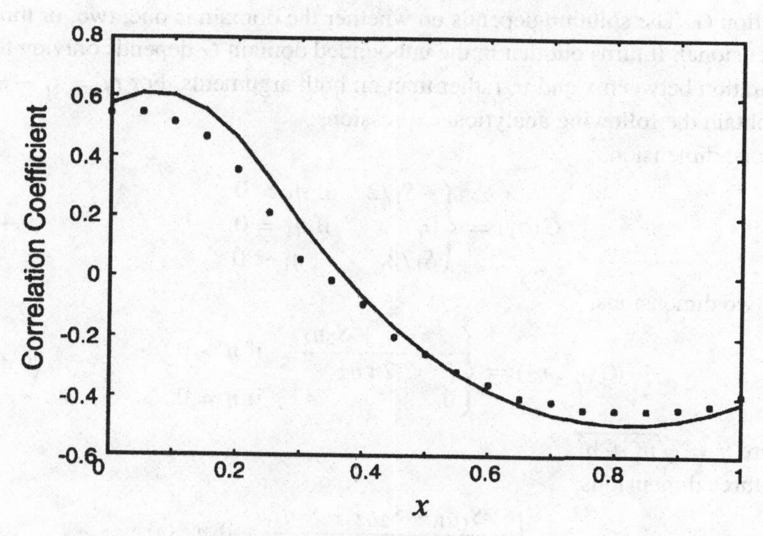

Figure 9.19 Comparison of $\phi(0.25)$ to $Y(x)$ correlation ($\sigma^2 = 2$).

9.5 Large-domain analytical solutions

Large-domain analysis is appropriate when the actual boundaries of the formation are far away from the area of interest or if we are uncertain about the location and effects of the boundaries. Because of its complexity, you may choose to skip this section at first reading.

We assume steady flow in an n-dimensional unbounded domain, with no sources or sinks; log-conductivity (or log-transmissivity) with constant mean; and head with constant mean slope S_i in direction i. For example, for three dimensions,

$$F = E[Y] = C_1 \tag{9.41}$$

$$H(x_1, x_2, x_3) = S_1 x_1 + S_2 x_2 + S_3 x_3 + C_2, \tag{9.42}$$

where S_1, S_2, and S_3 are known constants and C_1 and C_2 are unspecified constants. S_i is the slope of the mean piezometric head in the x_i direction. The equations satisfied by H and h are simplified as follows:

$$\frac{\partial^2 H}{\partial x_1^2} + \frac{\partial^2 H}{\partial x_2^2} + \frac{\partial^2 H}{\partial x_3^2} = 0, \tag{9.43}$$

which is satisfied by the solution, and

$$\frac{\partial^2 h}{\partial x_1^2} + \frac{\partial^2 h}{\partial x_2^2} + \frac{\partial^2 h}{\partial x_3^2} = -S_1 \frac{\partial f}{\partial x_1} - S_2 \frac{\partial f}{\partial x_2} - S_3 \frac{\partial f}{\partial x_3}. \tag{9.44}$$

Under these conditions, a general expression can be obtained for the influence function G. The solution depends on whether the domain is one, two, or three dimensional. It turns out that in the unbounded domain G depends only on the separation between $\mathbf{x}$ and $\mathbf{u}$, rather than on both arguments. For $\eta_i = x_i - u_i$, we obtain the following analytical expressions:
For one dimension,

$$G(\eta_1) = \begin{cases} -S_1/2, & \text{if } \eta_1 > 0 \\ 0, & \text{if } \eta_1 = 0. \\ S_1/2, & \eta_1 < 0 \end{cases} \tag{9.45}$$

For two dimensions,

$$G(\eta_1, \eta_2) = \begin{cases} -\dfrac{S_1\eta_1 + S_2\eta_2}{2\pi\eta^2}, & \text{if } \eta > 0 \\ 0, & \text{if } \eta = 0 \end{cases}, \tag{9.46}$$

where $\eta = \sqrt{\eta_1^2 + \eta_2^2}$.
For three dimensions,

$$G(\eta_1, \eta_2, \eta_3) = \begin{cases} -\dfrac{S_1\eta_1 + S_2\eta_2 + S_3\eta_3}{4\pi\eta^3}, & \text{if } \eta > 0 \\ 0, & \text{if } \eta = 0 \end{cases}, \tag{9.47}$$

where $\eta = \sqrt{\eta_1^2 + \eta_2^2 + \eta_3^2}$.

Basically, these results mean that an increase in log-conductivity at location $\mathbf{u}$ translates into an increase in head downgradient and a decrease upgradient of $\mathbf{u}$. The head fluctuation at location $\mathbf{x}$ depends only on the distance $\mathbf{x} - \mathbf{u}$ from the log-conductivity fluctuation and is given by the functions above.

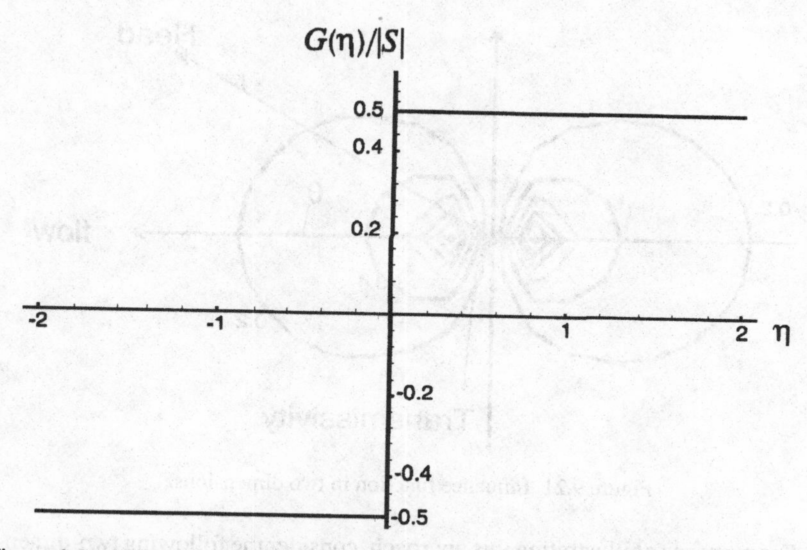

Figure 9.20 Influence function for one dimension. Positive η indicates that the head fluctuation is downgradient of the log-conductivity perturbation.

Note that the influence of a small fluctuation of log-conductivity on head does not decrease with distance for one-dimensional flow. See Figure 9.20, which shows $G(\eta)$ for $S_1 = -1$ (*i.e.*, the flow is in the positive direction). However, the influence decreases with the inverse of the distance for two-dimensional flow and with the inverse of the square distance for three-dimensional flow.

Let us examine the two-dimensional case in some detail. Equation (9.46) can be written

$$G(r, \theta) = \begin{cases} |S| \cos(\theta)]/[2\pi r], & \text{if } r > 0 \\ 0, & \text{if } r = 0 \end{cases}, \tag{9.48}$$

where r is the distance and θ is the angle with the direction of flow as shown in Figure 9.21. (This expression is the "dipole" of potential flow theory.) With the exception of the singularity at 0, the influence of a log-transmissivity fluctuation at a point at distance r is inversely proportional to r and is proportional to the absolute value of the head gradient. The influence is maximum in the direction of flow and zero at the direction exactly perpendicular to that of flow. The influence is positive downgradient and negative upgradient of the log-transmissivity fluctuation.

In interpreting these results, keep in mind that a very large domain is assumed so that the effects of the boundaries are neglected. Furthermore, the influence is computed for a *small* fluctuation of the log-transmissivity about a background of constant log-transmissivity.

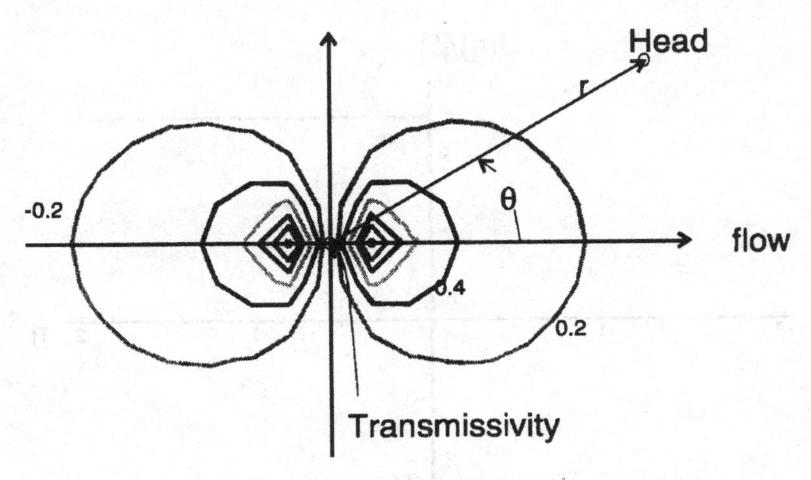

Figure 9.21 Influence function in two dimensions.

For purposes of illustrating this approach, consider the following two-dimensional case (adapted after Dagan [33]). Assume that the log-transmissivity Y is an isotropic stationary function with mean F and exponential covariance function

$$R_{YY}(\mathbf{x}, \mathbf{x}') = \sigma^2 \exp\left(-\frac{|\mathbf{x} - \mathbf{x}'|}{l}\right), \tag{9.49}$$

where σ^2 is the variance and ℓ is the integral scale of Y. With these assumptions Dagan found, for $S_1 = S$, $S_2 = 0$,

$$R_{\phi Y}(\mathbf{x}', \mathbf{x}) = \sigma^2 S\, \ell \frac{r_1}{r^2}[1 - (1 + r)\exp(-r)], \tag{9.50}$$

where

$$\mathbf{r} = (r_1, r_2) = \left(\frac{x_1 - x_1'}{l}, \frac{x_2 - x_2'}{l}\right), \quad r = \sqrt{r_1^2 + r_2^2}.$$

Note that the cross-covariance is positive when the head is downgradient of the log-transmissivity. The head covariance does not exist because in an unbounded domain the variance of the head can be infinitely large. However, the variogram can be defined:

$$\gamma_{\phi\phi}(\mathbf{x}, \mathbf{x}') = \frac{1}{2}\sigma^2 S^2 \ell^2 \left\{ \left(2\frac{r_1^2}{r^2} - 1\right)\left[\frac{1}{2} + \frac{\exp(-r)(r^2 + 3r + 3) - 3}{r^2}\right] + E_1(r) + \ln(r) + \exp(-r) - 0.4228 \right\}, \quad \text{if } r > 0 \tag{9.51}$$

$$\gamma_{\phi\phi}(r) = 0, \quad \text{if } r = 0, \tag{9.52}$$

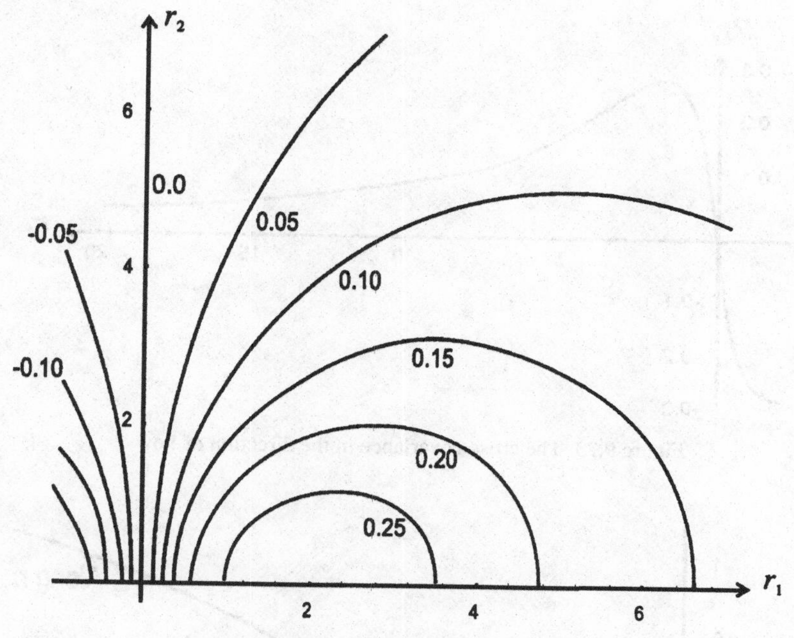

Figure 9.22 Contour plot of cross-covariance.

where $E_1(r) = \int_1^\infty \frac{\exp(-rt)}{t} dt$ is a tabulated function known as the exponential integral and C is a constant. (For convenience in computing function $E_1(r)$, MATLAB program `expint1` is given in Section 9.9.) At large separation distances, the variogram increases logarithmically with the distance. In particular, the variograms in the principal directions at large distances are

$$\gamma_{\phi\phi}(r_1) \approx \ln(r_1)/2 + 0.0386, \quad \gamma_{\phi\phi}(r_2) \approx \ln(r_2)/2 - 0.4613.$$

Near the origin, the variogram has parabolic behavior. Note that the variogram increases faster with the distance in the direction of flow, indicating that a higher degree of spatial continuity is found in the direction perpendicular to the flow.

Results are plotted for $\sigma^2 = 1$, $S = 1$, and $l = 1$. Figure 9.22 is a contour map of the cross-covariance of the log-transmissivity and head. Figure 9.23 shows the cross-covariance as a function of r_1, when $r_2 = 0$. It is worth noticing that the cross-covariance is an odd function in the direction of flow and an even function in the direction perpendicular to flow.

Figure 9.24 shows the (semi)variogram in the two principal directions: (1) along the flow and (2) across the flow. The variogram increases faster in the direction of flow, indicating that a head observation will be more informative

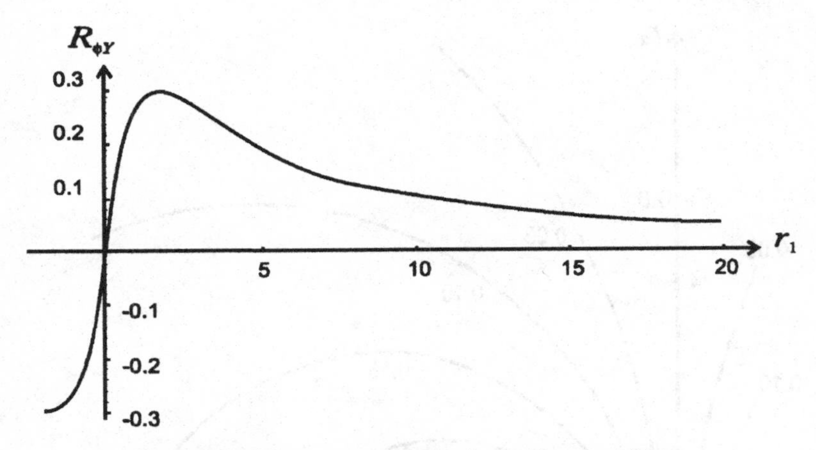

Figure 9.23 The cross-covariance in the direction of flow.

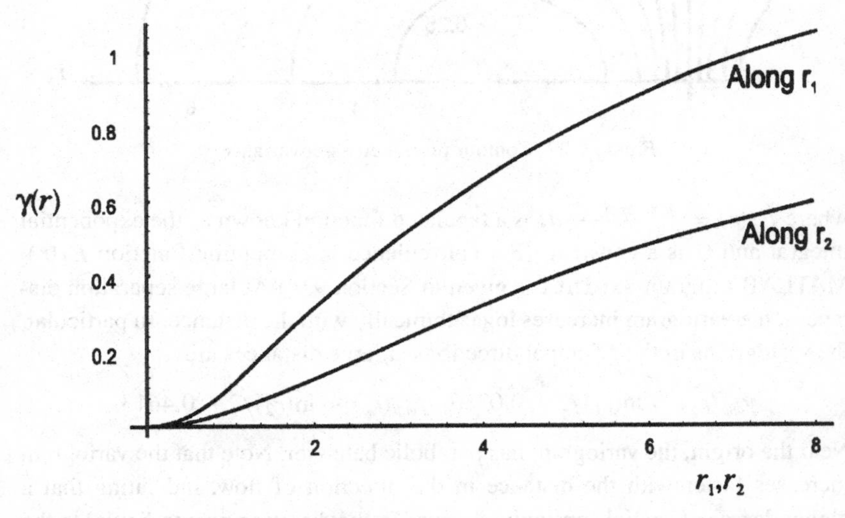

Figure 9.24 $\gamma_{\phi\phi}(r)$ in the two principal directions. r_1 is the distance in the direction of flow.

about the hydraulic head in the direction perpendicular to flow. Figure 9.25 shows a contour plot of the variogram.

In cokriging it is more convenient to work with the generalized covariance of the hydraulic head. Let

$$K_{\phi\phi}(r) = -\gamma_{\phi\phi}(r) \tag{9.53}$$

be a generalized covariance function that becomes 0 when $r = 0$.

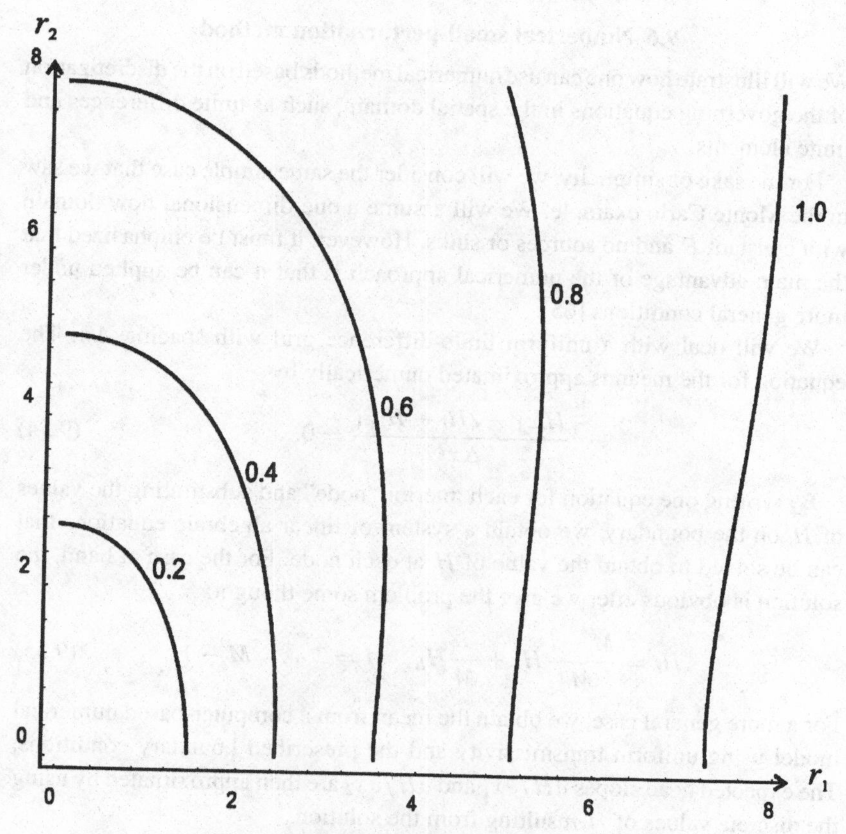

Figure 9.25 Contour plot of the variogram.

A useful closed-form analytical solution was thus obtained by making a number of simplifications about the form of the mean and the covariance function of $Y(x)$, the geometry of the flow domain, and the boundary conditions. The assumption of an unbounded domain is appropriate if the boundary of the domain is not well defined and values of the head or discharge on the boundary are not actually measured. This is often the case in regional flow modeling, where the only available information is measurements of head and transmissivity at interior points. Furthermore, if the objective of second moment characterization of ϕ is cokriging, approximate solutions may suffice given the relative insensitivity of linear estimation results to the form of the second moments. It has been shown [42] that the general characteristics of the analytical solution just described were found even in a finite domain that did not meet the mathematical assumptions of the analysis.

9.6 Numerical small-perturbation method

We will illustrate how one can use numerical methods based on the discretization of the governing equations in the spatial domain, such as finite differences and finite elements.

For the sake of simplicity, we will consider the same simple case that we saw in the Monte Carlo example. We will assume a one-dimensional flow domain with constant F and no sources or sinks. However, it must be emphasized that the main advantage of the numerical approach is that it can be applied under more general conditions [65].

We will deal with a uniform finite-difference grid with spacing Δx. The equation for the mean is approximated numerically by

$$\frac{H_{i-1} - 2H_i + H_{i+1}}{\Delta x^2} = 0. \tag{9.54}$$

By writing one equation for each interior "node" and substituting the values of H on the boundary, we obtain a system of linear algebraic equations that can be solved to obtain the value of H at each node. For the case at hand, the solution is obvious after we give the problem some thought:

$$H_i = \frac{M-i}{M} H_1 + \frac{i}{M} H_L, \quad i = 2, \ldots, M-1. \tag{9.55}$$

For a more general case, we obtain the mean from a computer-based numerical model using uniform transmissivity and the prescribed boundary conditions. The expected head slopes $\partial H / \partial x_1$ and $\partial H / \partial x_2$ are then approximated by using the discrete values of H resulting from the solution.

The stochastic partial differential equation for h, Equation (9.26), can also be written in a finite-difference form:

$$\frac{h_{i-1} - 2h_i + h_{i+1}}{\Delta x^2} = -\frac{f_{j+1} - f_{j-1}}{2\Delta x} \frac{\Delta H}{L} \tag{9.56}$$

or

$$h_{i-1} - 2h_i + h_{i+1} = -\frac{\Delta x}{2} \frac{\Delta H}{L} (f_{i+1} - f_{i-1}). \tag{9.57}$$

This equation is written for each node. In matrix form, this set of equations is

$$\mathbf{Ah} = \mathbf{Bf}, \tag{9.58}$$

where $\mathbf{h}$ is the vector of values of h sampled on the nodes, $\mathbf{f}$ is the same type of vector for f, and $\mathbf{A}$ and $\mathbf{B}$ are constant matrices. The same can be written as

$$\mathbf{h} = \mathbf{Cf}, \tag{9.59}$$

where $\mathbf{C} = \mathbf{A}^{-1}\mathbf{B}$.

Equation (9.59) is but the discrete equivalent of (9.28). It can be used to develop the required second moments:

$$E[\mathbf{h}\mathbf{h}^T] = \mathbf{C}E[\mathbf{f}\mathbf{f}^T]\mathbf{C}^T \tag{9.60}$$

$$E[\mathbf{h}\mathbf{f}^T] = \mathbf{C}E[\mathbf{f}\mathbf{f}^T]. \tag{9.61}$$

***Example 9.2** Let us continue with the one-dimensional flow example that we saw in Monte Carlo simulation. The* **h** *and* **f** *vectors of random variables are*

$$\mathbf{h} = \begin{bmatrix} h_1 \\ h_2 \\ \vdots \\ h_M \end{bmatrix}, \ \mathbf{f} = \begin{bmatrix} f_1 \\ f_2 \\ \vdots \\ f_M \end{bmatrix}.$$

The $M \times M$ **A** *matrix is*

$$\mathbf{A} = \begin{bmatrix} 1 & 0 & 0 & \cdots & 0 & 0 & 0 \\ 1 & -2 & 1 & \cdots & 0 & 0 & 0 \\ 0 & 1 & -2 & \cdots & 0 & 0 & 0 \\ & & & \ddots & & & \\ 0 & 0 & 0 & \cdots & 1 & -2 & 1 \\ 0 & 0 & 0 & \cdots & 0 & 0 & 1 \end{bmatrix}, \tag{9.62}$$

and the **B** *matrix is*

$$\mathbf{B} = -\frac{\Delta x}{2}\frac{\Delta H}{L} \begin{bmatrix} 0 & 0 & 0 & \cdots & 0 & 0 & 0 \\ 1 & 0 & -1 & \cdots & 0 & 0 & 0 \\ 0 & 1 & 0 & \cdots & 0 & 0 & 0 \\ & & & \ddots & & & \\ 0 & 0 & 0 & \cdots & 1 & 0 & -1 \\ 0 & 0 & 0 & \cdots & 0 & 0 & 0 \end{bmatrix}. \tag{9.63}$$

A comparison between the numerical and the analytical small-perturbation solution indicates that they are indistinguishable for the case depicted in Figures 9.10–9.19.

***Exercise 9.2** In the discrete formulation of numerical models, function $f(\mathbf{x})$ is replaced by vector* **f** *and function $h(\mathbf{x})$ is replaced by vector* **h**. *Identify the equivalents of the influence function $G(\mathbf{x}, \mathbf{u})$ and of Equations (9.28), (9.30), and (9.32) in the discrete formulation.*

9.7 An inverse problem

We will illustrate some of the concepts and methods we have covered through an application to an inverse problem. Consider steady two-dimensional flow in an unbounded domain with negligible pumping or recharge. The data consist of 9 observations of the transmissivity (see Table 9.1) and 19 observations of the hydraulic head (Table 9.2). The problem will be formulated using the linear estimation methods we have seen in this book.[1]

The transmissivity is measured in 10^{-4} m^2/sec, the head is in meters, and x and y are in units of 100 meters. The objective is to estimate the log-transmis-

Table 9.1. *Measured transmissivity and coordinates of observation points*

T	x	y
3.084	−6.0	4.0
0.442	−4.0	−3.0
0.263	−3.0	−4.0
0.340	−3.0	−2.0
1.022	−1.0	6.0
1.012	3.0	−3.0
0.526	5.0	−4.0
0.537	5.0	−3.0
0.278	7.0	−5.0

Table 9.2. *Measured head and coordinates of observation points*

ϕ	x	y	ϕ	x	y
13.832	−4.0	−5.0	9.113	0.0	7.0
12.406	−4.0	7.0	6.997	1.0	−2.0
11.809	−3.0	3.0	6.055	3.0	−6.0
10.830	−2.0	−5.0	5.593	3.0	4.0
10.605	−2.0	−2.0	4.924	4.0	1.0
9.143	−1.0	−4.0	4.163	5.0	−2.0
9.644	−1.0	1.0	2.971	5.0	5.0
10.391	−1.0	6.0	2.569	6.0	−7.0
8.980	0.0	2.0	2.834	6.0	−3.0
9.542	0.0	5.0			

[1] The method assumes that the transmissivity varies gradually and that the linearization about the prior mean is adequate [34]. For methods that iterate in order to linearize about the conditional mean and that could be useful in large-variance cases, see [19, 85, and 148]. There are also other nonlinear estimation methods, such as the one in reference [22].

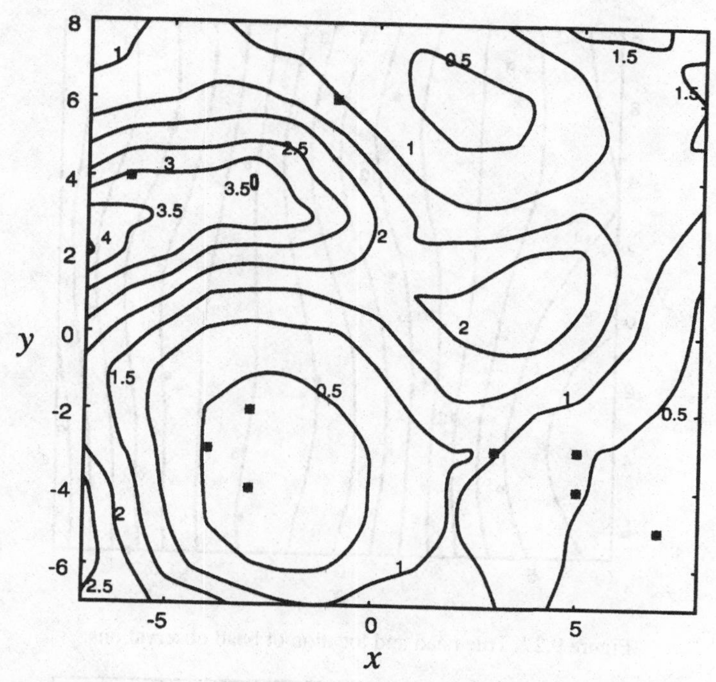

Figure 9.26 True transmissivity and location of transmissivity observations.

sivity and the hydraulic head in the domain $x = [-7, 8]$, $y = [-7, 8]$. Figure 9.26 shows the location of the transmissivity observations and the contour map (or lines of equal value) of the true transmissivity. For instructional purposes, we consider a synthetic case where the transmissivity is known although only some measurements are used in estimation. We do this so that we can evaluate the effectiveness of the estimation method at the end of the exercise. Figure 9.27 shows the locations of the head observations and the contour map of the true head.

We may start by using only the transmissivity data to perform kriging. It is common practice to work with and apply kriging on $Y = \ln T$. Because we have only 9 observations, we fit a simple variogram to the log-transmissivity data:

$$\gamma = 0.1\, h, \tag{9.64}$$

where h is the separation distance. Then, we estimate through kriging the best estimate of log-transmissivity $\hat{Y}$ and we plot $\exp(\hat{Y})$ in Figure 9.28. The estimated transmissivity[2] is a much smoother and simpler function than the actual

[2] If $\hat{Y}$ is the conditional mean of Y, $\exp(\hat{Y})$ is not the conditional mean of T. Nevertheless, $\exp(\hat{Y})$ is median value of T and in this sense is a reasonable estimate of T.

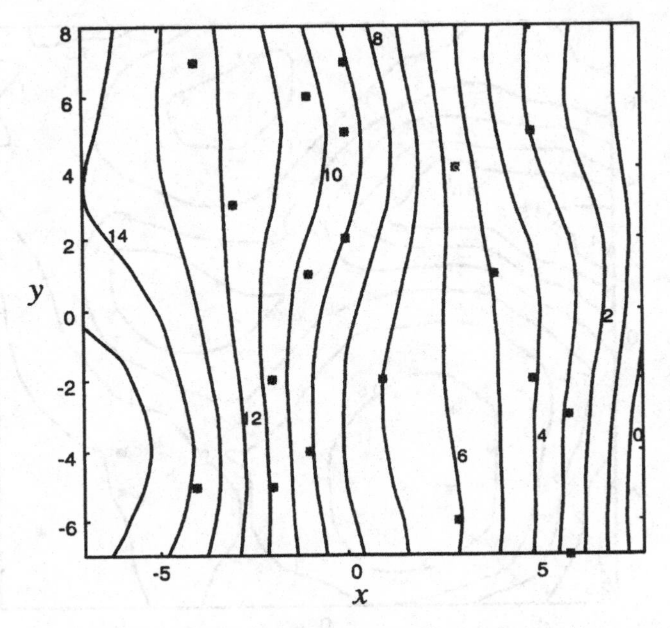

Figure 9.27 True head and location of head observations.

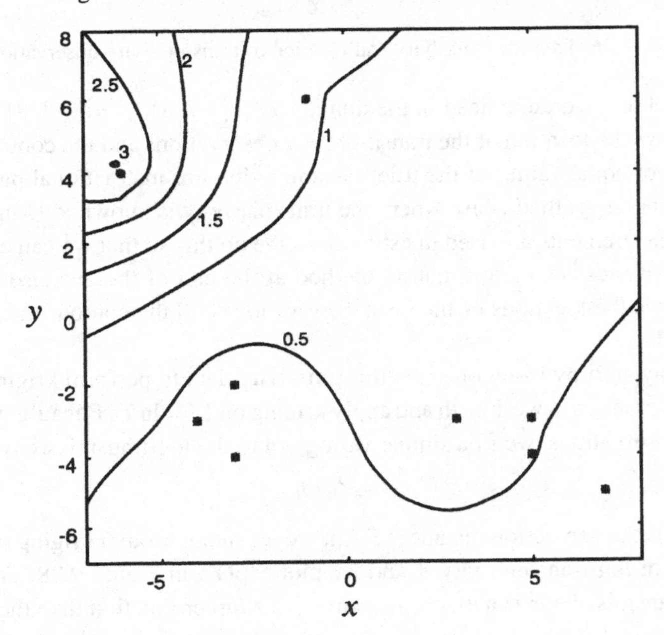

Figure 9.28 Location of transmissivity data and estimate of transmissivity obtained using only these data.

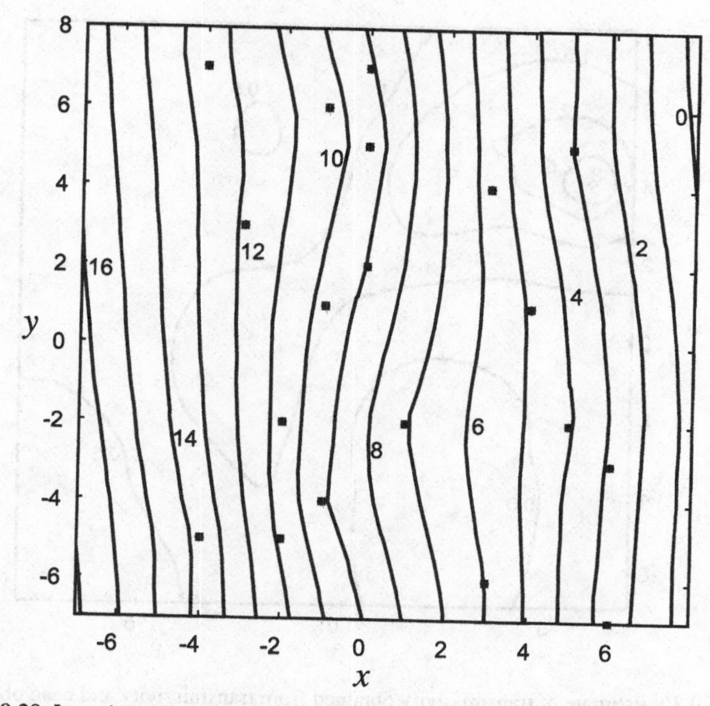

Figure 9.29 Location of head data and estimate of head obtained using only these data.

one. The transmissivity data allowed us to identify some coarse features of the actual function but because the observations are few and nonuniformly distributed they cannot give any information about the function at the northeast of the domain and elsewhere. Similarly, we analyze the head data to establish the following simple model. The mean is variable,

$$H(x, y) = S_0 + S_1 x + S_2 y, \tag{9.65}$$

where S_0, S_1, and S_2 are three unknown drift coefficients, and the variogram γ (or generalized covariance function K) is linear,

$$\gamma(h) = 0.077h, \quad K(h) = -0.077h. \tag{9.66}$$

The best estimates obtained using (universal) kriging with this model are shown in Figure 9.29. The estimates are clearly quite accurate where measurements abound.

An estimate of the transmissivity using cokriging, using both transmissivity and head observations, is presented in Figure 9.30. It is obvious that many more details of the actual transmissivity were identified using the head data in addition to the transmissivity data. Although many small-scale features remain

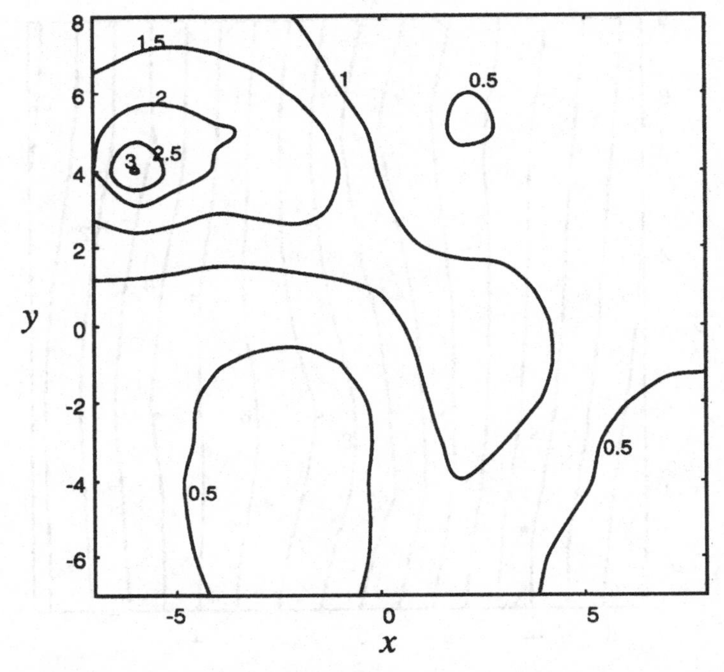

Figure 9.30 Estimate of transmissivity obtained from transmissivity and head observations.

undetected most of the important large-scale features of T have been determined. The same cokriging approach was applied to obtain best estimates of the head (shown in Figure 9.31). The estimate of the head in most of the domain is practically perfect. In the remainder of this section, we will provide some more details on how these estimates were obtained.

We will use cokriging as given in Chapter 8 and will derive second moments of head following the analytic small-perturbation approach, described in Section 9.5. The log-transmissivity is modeled as stationary with unknown mean F and covariance function

$$R_{YY}(\mathbf{x}_i - \mathbf{x}_j) = \sigma^2 \exp\left(-\frac{\|\mathbf{x}_i - \mathbf{x}_j\|}{l}\right), \tag{9.67}$$

which has two parameters, σ^2 and l. The mean head is modeled as having variable mean (Equation (9.65)), where S_0 is treated as unknown whereas S_1 and S_2 are treated as known, consistent with the analysis of Section 9.5. From Figure 9.29, we roughly estimate that $S_1 = -1$ and $S_2 = 0$. As discussed in Chapter 8, we still need to define the head generalized covariance $K_{\phi\phi}$ and the head–log-transmissivity cross-covariance $R_{\phi Y}$ to be able to solve the cokriging

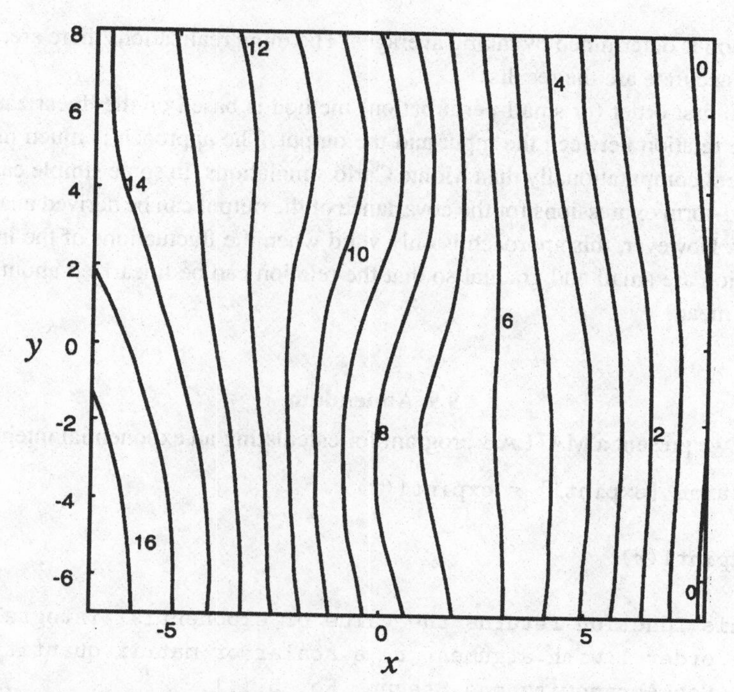

Figure 9.31 Estimate of head obtained from transmissivity and head observations.

problem. The generalized covariance is found from Equations (9.53) and (9.51), using $S = -1$. The cross-covariance is given from Equation (9.50). Thus, we may proceed to apply cokriging.

Exercise 9.3 *Write out the cokriging equations for estimating log-transmissivity and head under the assumptions of this section.*

9.8 Key points of Chapter 9

In many important applications, different spatial functions are related by the equations that describe the flow and transport in geologic formations. These equations may be the best way to determine the correlation structure of a dependent variable, such as head, from the correlation structure of an independent variable, such as conductivity. Two possible methods are Monte Carlo simulations and first-order analysis.

Monte Carlo simulations are based on the mathematical generation of many input-function realizations and the solution of the governing equations to compute output-function realizations. Then the correlation structure of the output-function is determined by taking averages. The more realizations there are, the more accurate are the results.

The first-order (or small-perturbation) method is based on the linearization of the relation between the input and the output. The approach is much more efficient computationally than Monte Carlo simulations. In some simple cases, closed-form expressions for the covariance of the output can be derived analytically. However, this approach is only valid when the fluctuations of the input function are small and gradual so that the relation can be linearized about the prior mean.

9.9 Addendum

Here we present a MATLAB program for calculating an exponential integral:

```
function [expint1] = expint1(r)
%
% expint1(r)
%
% This function returns the value of exponential integral
%    order 1 with argument r, a scalar or matrix quantity.
%    See Abramowitz and Stegun, Eq. 5.1.1.
%
if r<1
%
a0 = -0.57721566;
a1 = 0.99999193;
a2 = -0.24991055;
a3 = 0.05519986;
a4 = -0.00976004;
a5 = 0.00107857;
expint1 = -log(r)+a0+a1*r+a2*r.^2+a3*r.^3+a4*r.^4+a5*r.^5;
else
a1 = 2.334733;
a2 = 0.250621;
b1 = 3.330657;
b2 = 1.681534;
expint1 = exp(-r).*(r.^2+a1*r+a2)./(r.^3+b1*r.^2+b2*r);
end
```

For example, the command `expint1(0.5)` should return the value 0.5598. At the limit as $r \to 0$, $E_1(r) \to +\infty$ but $E_1(r) + \ln(r) \to -0.5772$.

Appendix A
Probability theory review

In this appendix we review some basic concepts and results of probability theory that will be of use in this book. It is meant to be a refresher of concepts that the reader has already seen elsewhere in a more complete form and at the same time as a way to establish notation. The choice of topics reflects the emphasis of this book on using mean values, variances, and covariances. Useful textbooks, which could be used in a formal course on probability theory, are listed at the end of this appendix.

A.1 Introduction

A.1.1 Experiments

Probability theory is concerned with "experiments" with multiple possible outcomes. An example of an experiment is to toss a coin. The only outcomes (also known as realizations or sample values) are "heads" and "tails." We say that the ensemble of all possible realizations of this experiment are heads and tails. Another example, with many possible outcomes, is to spin a wheel of fortune.

Consider now the problem of predicting the piezometric head of an aquifer at a location. If we cannot predict the outcome but we realize that there are many possible outcomes (predictions), then it is useful to conceptualize the process the same way we do a game of chance. Prediction with incomplete information and games of chance share the characteristic that their outcomes cannot be predicted with certainty and we have to come up with a different (from the usual deterministic) way to make predictions. In some instances, we may be able to evaluate the likelihood of an "event," such as that the head is less than 420 meters above mean sea level. Although this is perhaps not as satisfactory for general purposes as predicting the exact value of the head (which is not possible given the available information), it is information useful to someone faced with the decision whether to drill a well at that location.

A.1.2 Probability

Intuitively, one may define the probability of an event as the number between zero and one that represents the chance or relative frequency of occurrence of the event. The probabilities of all possible (mutually exclusive) events of an experiment must sum to 1.

A.1.3 Random variables

In practically all cases we deal with experiments whose outcomes can be described through numerical values. The (uncertain) outcome of such an experiment is called a random variable. Unlike a deterministic variable, which has a unique value, a random variable may take any of many values, depending on the outcome of an experiment. A deterministic variable is characterized (fully described) by its unique numerical value (and, of course, its units). For example, the transmissivity is 1,000 m^2/day. By contrast, a random variable is characterized by a *probability distribution* that can be used to assign probabilities to any possible outcome. For example, consider the experiment of tossing a coin. Assign to the outcome a numerical value x, with $x = 1$ if heads and $x = 0$ if tails. (These are the only two possible outcomes of the experiment.) The random variable we have just defined is characterized as soon as we define the probability of all outcomes, *e.g.*, $\Pr[x = 0] = \Pr[x = 1] = 1/2$. The random variable may be denoted by an upper-case letter X so that it can be distinguished from the outcome of the experiment, which is denoted by a lower-case letter. (However, this convention is not maintained when the distinction is obvious.)

When the random variable can take only the discrete values $x_1, \ldots, x_n$, the probability law is equivalent to defining the probability of each outcome,

$$P_i = \Pr[X = x_i], \quad i = 1, \ldots, n. \tag{A.1}$$

The set of values $(P_1, \ldots, P_n)$ constitutes a discrete probability distribution. In most cases of practical value we are interested in experiments whose outcomes can take any real value in some interval(s). To describe the probability law of such a continuous random variable, we may use the *cumulative distribution function* (cdf) defined as follows:

$$F_X(x) = \Pr[X \leq x]. \tag{A.2}$$

For example, if $F_X(x) = 1 - \exp(-x)$, for $x \geq 0$, then $F_X(1) = 0.623$ means that the probability that the outcome of the experiment (realization of the random variable) will take values smaller than 1 is equal to 0.623. Note that since $F(x)$ is a probability, it must be nonnegative, and the probability of the "certain" event (*i.e.*, $X \leq \infty$) must be equal to one.

For a continuous random variable it is convenient to define the density of its probability, called a *probability density function* or pdf, denoted by $f_X(x)$, and defined in an indirect way as follows:

$$\int_{x_1}^{x_2} f_X(u)du = \Pr[x_1 \leq X \leq x_2] \tag{A.3}$$

or, for an infinitesimal dx,

$$f_X(x)dx = \Pr[x \leq X \leq x + dx]. \tag{A.4}$$

The cumulative distribution function and the probability density function are obviously related through

$$F_X(x) = \Pr[-\infty \leq X \leq x] = \int_{-\infty}^{x} f_X(u)du \tag{A.5}$$

or

$$F_X(x_2) - F_X(x_1) = \int_{x_1}^{x_2} f_X(u)du \tag{A.6}$$

and

$$\frac{dF_X}{dx} = f_X(x). \tag{A.7}$$

That is, the pdf of a random variable is the derivative of the cdf with respect to the argument x.

The pdf $f_X(x)$ has as units the inverse of the units of x. It must satisfy the following properties:

$$f_X(x) \geq 0 \tag{A.8}$$

$$\int_{-\infty}^{+\infty} f_X(u)du = 1. \tag{A.9}$$

A.2 Jointly distributed random variables

In many applications, we are interested in the joint behavior of two random variables. Consider, for example, the value of transmissivity at two nearby locations in the same aquifer. The two locations are aligned downgradient of a contaminant source. To evaluate the chance of a low resistance pathway, we would like to quantify the probability that both transmissivities are high. Furthermore, we would like to evaluate how measuring the one transmissivity affects the probability that the other transmissivity is high. To answer these questions we need to define the joint probability law of two or more random variables. We will restrict our attention to continuous random variables.

To keep the notation simple let us consider only two random variables, X and Y. The joint cumulative distribution function is defined by

$$F_{X,Y}(x, y) = \Pr[X \leq x \quad \text{and} \quad Y \leq y]. \tag{A.10}$$

Again, the capital letters X and Y stand for the random variables and the lowercase letters x and y are realizations or sample values (variables taking numerical values).

The joint probability density function $f_{X,Y}(u, v)$ is defined through the property that its integral over a domain gives the probability that the outcome will be in this domain. That is

$$\int_D \int f_{X,Y}(u, v) du\, dv = \Pr[(X, Y) \in D], \tag{A.11}$$

where D is a domain in the space where X and Y take values (e.g., D is the domain of nonnegative outcomes of X and Y) and $\in$ stands for "belongs to." Or, for infinitesimally small dx and dy,

$$f_{X,Y}(x, y) dx\, dy = \Pr[x \leq X \leq x + dx \quad \text{and} \quad y \leq Y \leq y + dy]. \tag{A.12}$$

Obviously,

$$f_{X,Y}(x, y) \geq 0 \tag{A.13}$$

and

$$\int_{-\infty}^{+\infty} \int_{-\infty}^{+\infty} f_{X,Y}(u, v) du\, dv = 1. \tag{A.14}$$

We define the *marginal* pdf of x as

$$f_X(x) = \int_{-\infty}^{+\infty} f_{X,Y}(x, u) du. \tag{A.15}$$

Similarly, the marginal cdf is the already familiar cdf of random variable X. The same holds true for the marginal functions of y. In a sense, the marginal functions allow us to concentrate on the probability law of one variable by averaging out the variability of the other.

The *conditional* pdf of X, given $Y = y$, is denoted by $f(x \mid y)$ and is defined as the ratio of the joint of X and Y over the marginal of Y,

$$f_{X|Y}(x \mid y) = \frac{f_{X,Y}(x, y)}{f_Y(y)}. \tag{A.16}$$

This can be interpreted as the probability density function of random variable X conditional on the information that the outcome of the other is $Y = y$. For example, let X and Y denote the values of transmissivity at two locations in close proximity to each other. Before any measurements are made, the probability law of X is given by the marginal pdf $f_X(x)$. However, once Y is measured and found to be equal to y, the probability law of X is given by $f_{X|Y}(x \mid y)$. Thus, a high value of y may increase the probability that X is also high.

In the special case that the marginal is equal to the conditional, *i.e.*,

$$f_{X|Y}(x \mid y) = f_X(x), \tag{A.17}$$

the two random variables are called (in a probabilistic sense) *independent*. It follows from Equations (A.16) and (A.17) that for independent random variables the joint pdf is equal to the product of the marginal pdfs:

$$f_{X,Y}(x, y) = f_X(x) f_Y(y). \tag{A.18}$$

For independent random variables, information about the outcome of the one does not affect the probability of the outcome of the other.

A.3 Expectation and moments

If x is the realization of a random variable with discrete probability distribution $(P_1, \ldots, P_n)$ and $g(x)$ is a deterministic function, *e.g.*, $g(x) = x^2$, then the expected value of g is a deterministic quantity defined as follows:

$$E[g(X)] = \sum_{i=1}^{n} g(x_i) P_i. \tag{A.19}$$

The expected value thus represents the average of all possible outcomes of $g(X)$ weighted by the probability of their occurrence. Similarly, in the case of a continuous random variable, the expected value is

$$E[g(X)] = \int_{-\infty}^{+\infty} g(x) f_X(x) dx. \tag{A.20}$$

The process of calculating the expected value is linear (like summation or integration) and is called the *expectation operator* or *ensemble averaging*. We will only consider the continuous case from this point.

Of particular interest is the *mean* of a random variable defined as

$$\mu = E[X] = \int_{-\infty}^{+\infty} x \, f_X(x) dx. \tag{A.21}$$

The mean is also known as the first statistical *moment*. The i-th *moment about the origin* is defined as

$$m_i = \int_{-\infty}^{+\infty} x^i f_X(x) dx, \tag{A.22}$$

where i is a positive integer. Also of interest are the *central* (*i.e.*, about the mean value) *moments*. The i-th central moment is defined as follows:

$$M_i = \int_{-\infty}^{+\infty} (x - \mu)^i f_X(x) dx. \tag{A.23}$$

The second central moment is commonly used as a measure of the spread of the distribution about the mean. It is called the *variance* and is usually denoted by σ^2. Its square root is known as the *standard deviation*. (Note that a deterministic real number can be seen as a random variable with zero variance.) A useful relation is

$$E[X^2] = \mu^2 + \sigma^2. \tag{A.24}$$

The expectation operator can be defined for many random variables in a similar way. Consider, for example, the case of two random variables X and Y with joint pdf $f_{X,Y}(x, y)$ and a deterministic function $g(x, y)$. The expected value of g is defined as the weighted average,

$$E[g(X, Y)] = \int_{-\infty}^{+\infty} \int_{-\infty}^{+\infty} g(x, y) f_{X,Y}(x, y) dx\, dy. \tag{A.25}$$

We can also define moments about the origin,

$$m_{i,j} = \int_{-\infty}^{+\infty} \int_{-\infty}^{+\infty} x^i y^j f_{X,Y}(x, y) dx\, dy, \tag{A.26}$$

and about the mean,

$$M_{i,j} = \int_{-\infty}^{+\infty} \int_{-\infty}^{+\infty} (x - E[X])^i (y - E[Y])^j f_{X,Y}(x, y) dx\, dy. \tag{A.27}$$

Of particular interest is the central moment with $i = j = 1$, known as the covariance and denoted by c_{XY}:

$$c_{XY} = \int_{-\infty}^{+\infty} \int_{-\infty}^{+\infty} (x - E[X])(y - E[Y]) f_{X,Y}(x, y) dx\, dy. \tag{A.28}$$

A normalized and dimensionless form of the covariance is the correlation coefficient

$$\rho_{XY} = \frac{c_{XY}}{\sigma_X \sigma_Y}. \tag{A.29}$$

One can verify that ρ is dimensionless and can take values only in the interval $[-1, 1]$. If X and Y are independent random variables then the correlation coefficient is exactly zero. The opposite is not necessarily true, except in the important special case that the random variables have a Gaussian distribution (which we will discuss elsewhere). In linear estimation, which does not use complete distributions but implicitly assumes that a Gaussian distribution is not too far away from the truth, we treat the absolute value of the correlation coefficient as the measure of dependence between two variables. In reality, $|\rho_{XY}|$ measures the degree of *linear dependence* between two random variables. However, in linear estimation, when we talk of dependence we usually mean linear dependence. When $|\rho_{XY}| = 1$ the realizations of the random variables X and Y satisfy a linear relation

$$x = ay + b, \tag{A.30}$$

where a and b are constants and a has the same sign as ρ_{XY}. Values of $|\rho_{XY}|$ near one indicate a high degree of linear dependence, *i.e.*, Equation (A.30) is a good approximation. Values near zero indicate practically no linear dependence between x and y.

Moments are important because, among other applications, they are often used instead of complete distributions to characterize random variables. In linear geostatistics we deal, whenever possible, only with mean values, variances, and covariances.

A.4 Functions of random variables

Let $y = g(x)$ be a deterministic function. If x is a realization of a random variable X with given probability density function $f_X(x)$, then for every realization of X we can compute a value of y. These values of y are the realizations of another random variable Y. In applications we often are faced with the problem of *derived distributions*, *i.e.*, calculating the pdf of Y, $f_Y(y)$, from the pdf of X, $f_X(x)$. For example, we may have to determine the pdf of piezometric head from the pdf of transmissivity, where the two variables are related via continuity and Darcy's law.

The cumulative distribution function of Y is

$$F_Y(y) = \Pr[Y \leq y] = \Pr[g(X) \leq y]. \tag{A.31}$$

This relation can be used to derive $F_Y(y)$ and then $f_Y(y)$. However, with the exception of special cases, this method of calculation is fraught with difficulties. An analytical method applicable when g is differentiable and is continuous in an interval yields

$$f_Y(y) = \sum_k \frac{f_X(x_k)}{|g'(x_k)|}, \tag{A.32}$$

where x_k is a root of the equation $y - g(x) = 0$ and g' is the derivative of g. (If g is monotonic, then $k = 1$.)

These methods can be extended to cases of many input and output variables. Consider, for example, the case of two of each:

$$U = g_1(X, Y) \tag{A.33}$$

$$V = g_2(X, Y). \tag{A.34}$$

The problem is how to determine $f_{U,V}(u, v)$ from $f_{X,Y}(x, y)$. For differentiable g_1 and g_2 and continuous random variables, it can be computed analytically according to the relation

$$f_{U,V}(u, v) = \sum_k \frac{f_{X,Y}(x_k, y_k)}{|(J(x_k, y_k)|}, \tag{A.35}$$

where for given values of u and v, x_k and y_k are roots of the system of equations, and J is the Jacobian, which in this case is

$$J = \frac{\partial g_1}{\partial x}\frac{\partial g_2}{\partial y} - \frac{\partial g_1}{\partial y}\frac{\partial g_2}{\partial x}. \tag{A.36}$$

This method gives analytical closed-form solutions in, among other cases, the special case of Gaussian distributions and linear relations.

An approximate method that is widely used in applications is *first-order analysis*. For example, consider the case of an input variable and an output variable. First, we linearize the relation about the mean of the input variable:

$$g(x) \simeq g(\mu_X) + g'(\mu_X)(x - \mu_X), \tag{A.37}$$

where $g'(\mu_X)$ is the derivative of g evaluated at the mean of X. We then use (A.37) to calculate the mean and variance of Y and the covariance between X and Y:

$$\mu_Y \simeq g(\mu_X) \tag{A.38}$$

$$\sigma_Y^2 \simeq [g'(\mu_x)]^2 \sigma_X^2 \tag{A.39}$$

$$c_{XY} \simeq g'(\mu_x)\sigma_X^2. \tag{A.40}$$

In this fashion, the first two moments of the joint distribution are approximated.

A.5 Further reading

There are, of course, countless books on probability theory. I will only mention two of my favorites: Drake [45] is a complete first textbook in applied probability theory with many examples and interesting problems and brain twisters. Benjamin and Cornell [10] provide a good introduction to probabilistic methods with excellent examples from engineering.

A.6 Review exercises

Exercise A.1 *Prove that if two random variables are independent, then their correlation coefficient is exactly 0.*

Exercise A.2 *If X has mean μ and variance σ^2, then find the mean and variance of*

$$Y = aX + b,$$

where a and b are constants, and also find the correlation coefficient between X and Y.

Exercise A.3 *Consider random variable X with cdf given by*

$$F_X(x) = 1 - \exp\left(-\frac{x}{l}\right), \quad x \geq 0,$$

where l is positive. Find the pdf of X. Plot the pdf and the cdf for $l = 1$.

Exercise A.4 *Consider random variable X with standardized normal pdf of*

$$f_X(x) = \frac{1}{\sqrt{2\pi}} \exp\left(-\frac{x^2}{2}\right).$$

(a) Find the mean and the variance of X.
(b) Derive the pdf of

$$Y = aX + b.$$

Exercise A.5 *Consider the random variable X with a pdf of*

$$f_X(x) = \frac{1}{\sqrt{2\pi}} \exp\left(-\frac{x^2}{2}\right).$$

(a) Find the mean and variance of

$$Y = X^2.$$

(b) Find the correlation between X and Y.
(c) Derive the pdf of Y.

Exercise A.6 *Let X and Y be two jointly distributed random variables. The mean and variance of X are m_X and σ_X^2, the mean and variance of Y are m_Y and σ_Y^2 and the correlation coefficient between X and Y is ρ_{XY}.*

(a) What is the best (i.e., with minimum mean square error) estimate of Y? (This is known as the unconditional estimate.) Show that it is unbiased. What is the mean square error associated with this estimate?
(b) What is the best linear unbiased estimate of Y given the additional information that X was sampled and was found to be x. (This is known as the conditional estimate.) Show that it is unbiased. What is the mean square

error associated with this estimate? (Hint: The linear estimator in this case is $\hat{y} = \kappa + \lambda x$. That is, the estimate is a weighted average of the observation and of a constant, which represents information from the specification of the pdf. You have to determine the value of the two coefficients κ and λ by minimizing the mean square error.)

Exercise A.7 *Consider a tagged particle that has a probability $1 - p$ of being in the mobile phase (the water in connected pores where a particle moves with a certain average velocity) and p of being in the immobile phase of the porous medium (e.g., sorbed on the solids). Determine the retardation coefficient of the particle. (In groundwater studies, we use the retardation coefficient, which is the average velocity in the mobile phase divided by the average velocity of a tagged particle.)*

Exercise A.8 *A tagged particle is subject to Brownian motion with dispersion coefficient D in one dimension. That is, starting from some location at time $t = 0$, it moves according to the following equation:*

$$x(t_2) = x(t_1) + \sqrt{2D(t_2 - t_1)}\,\epsilon, \quad \text{where } t_2 \geq t_1,$$

where ϵ is a normal random variable with mean 0 and variance 1, independent of the location or the past path of the particle.

(a) Find the mean and variance of the location of the particle at time t.

(b) Write the probability distribution of $x(t)$. (Hint: You may use, if you want, the following facts from probability theory: A normal random variable with mean μ and variance σ^2 has a pdf of

$$\frac{1}{\sqrt{2\pi\sigma^2}} \exp\left(-\frac{(x-\mu)^2}{2\sigma^2}\right).$$

A random variable that can be expressed as a linear function of normally distributed random variables is also normally distributed.)

(c) Show that the pdf of the location of a tagged particle at time t, $p(x, t)$, satisfies the diffusion equation:

$$\frac{\partial p}{\partial t} = D\frac{\partial^2 p}{\partial x^2}.$$

(d) For $D = 0.0001\ m^2/day$, compute and plot the pdf of the location of a tagged particle at $t = 10$ days. Then, simulate on the computer the location of 1,000 particles at time $t = 10$ days, plot their histogram, and compare this with the pdf. (Hint: You may use program `randn` *in MATLAB.)*

(e) Calculate and plot the path of a tagged particle over 10 days using a time interval of $\Delta t = 0.05$ days.

Exercise A.9 *For the tagged particle that is subjected to Brownian motion of the previous problem: Show that for any two times t_1 and t_2, $t_2 \geq t_1 \geq 0$,*

$$E\left[x(t_2) - x(t_1)\right] = 0$$

and

$$\frac{1}{2}E[(x(t_2) - x(t_1))^2] = D(t_2 - t_1).$$

(a) What kind of variogram describes the Brownian motion? Discuss the properties of realizations of functions that have this property.

(b) Brownian motion is called a process with "independent" increments. That is, if we see the process as the sum of many consecutive steps, each step should take place independently of past steps. Justify this by arguing that for any $t_3 \geq t_2 \geq t_1$,

$$E[(x(t_2) - x(t_1))(x(t_3) - x(t_2))] = 0.$$

Appendix B
Lagrange multipliers

B.1 The method

In kriging, we select the weights $\lambda_1, \ldots, \lambda_n$ that minimize the expression for the mean square error while at the same time meeting the unbiasedness constraints. This problem has the general form:

$$\min f(\lambda_1, \ldots, \lambda_n) \tag{B.1}$$

subject to

$$g_k(\lambda_1, \ldots, \lambda_n) = b_k, \quad k = 1, \ldots, p. \tag{B.2}$$

For example, for ordinary kriging (Chapter 3) with stationary isotropic functions,

$$\begin{aligned}\min f(\lambda_1, \ldots, \lambda_n) = & \sum_{i=1}^{n}\sum_{j=1}^{n} \lambda_i \lambda_j R(|\mathbf{x}_i - \mathbf{x}_j|) \\ & - 2\sum_{i=1}^{n} \lambda_i R(|\mathbf{x}_i - \mathbf{x}_0|) + R(0)\end{aligned} \tag{B.3}$$

subject to the condition

$$\sum_{i=1}^{n} \lambda_i = 1. \tag{B.4}$$

Equations (B.1) and (B.2), which constitute a *constrained optimization* problem, can be solved through the *method of Lagrange multipliers*. The steps in this method are:

1. Form the *Lagrangian* $L(\lambda_1, \ldots, \lambda_n, \nu_1, \ldots, \nu_p)$

$$\begin{aligned}L(\lambda_1, \ldots, \lambda_n, \nu_1, \ldots, \nu_p) = & f(\lambda_1, \ldots, \lambda_n) + 2\nu_1(g_1(\lambda_1, \ldots, \lambda_n) - b_1) \\ & + \cdots + 2\nu_p(g_p(\lambda_1, \ldots, \lambda_n) - b_p),\end{aligned} \tag{B.5}$$

   where $\nu_1, \ldots, \nu_p$ are parameters called Lagrange multipliers. The 2 is used in kriging only for convenience.

2. Take derivatives of L with respect to $\lambda_1, \ldots, \lambda_n, \nu_1, \ldots, \nu_p$ and set them equal to zero:

$$\frac{\partial f}{\partial \lambda_i} + 2\nu_1 \frac{\partial g_1}{\partial \lambda_i} + \cdots + 2\nu_p \frac{\partial g_p}{\partial \lambda_i} = 0, \quad i = 1, \ldots, n \tag{B.6}$$

$$g_k = b_k, \quad k = 1, \ldots, p. \tag{B.7}$$

Equation (B.7) is the same as (B.2), the constraints. The meaning of Equation (B.6) will be explained later.

3. Solve the system of $n + p$ linear equations, Equations (B.6) and (B.7), with $n + p$ unknowns.

For more details on this method, see reference [92].

B.2 Explanation

We will explain how the method of Lagrange multipliers achieves the desired objective by focusing on a simple example: ordinary kriging, Equations (B.3) and (B.4) with only two measurements. The covariance function is $R(\mathbf{x}, \mathbf{x}') = \exp(-|\mathbf{x} - \mathbf{x}'|/5)$, and the locations are $x_1 = [^0_0]$, $x_2 = [^3_4]$, and $x_0 = [^3_0]$.

The objective is to minimize $f(\lambda_2, \lambda_2)$,

$$\min f(\lambda_1, \lambda_2) = \lambda_1^2 + \lambda_2^2 + 0.7358\lambda_1\lambda_2 - 1.0976\lambda_1 - 0.8987\lambda_2 + 1, \tag{B.8}$$

subject to

$$\lambda_1 + \lambda_2 = 1. \tag{B.9}$$

Figure B.1 shows a contour map of $f(\lambda_1, \lambda_2)$. (Note that we did not have to specify the values of the measurements!) The minimum is at the center of the ellipses, at $\lambda_1 = 0.4435$, $\lambda_2 = 0.2862$. It can be calculated algebraically by setting the derivatives of the mean square error, Equation (B.8), with respect to λ_1 and λ_2 equal to zero and then solving the system of linear equations. This method is well known from basic calculus.

However, this solution does not satisfy the constraint (B.9). The problem is to find the values of λ_1 and λ_2 that give the smallest possible value f while satisfying the constraint (*i.e.*, the values must be located on the straight line shown in Figure B.1). Graphically, this point can be found on this figure where the line of the constraint is tangent to a contour line of constant f.

The algebraic equivalent of this condition is at the heart of the method of *Lagrange multipliers*. At any (λ_1, λ_2), the vector

$$\nabla f = \begin{bmatrix} \frac{\partial f}{\partial \lambda_1} \\ \frac{\partial f}{\partial \lambda_2} \end{bmatrix}$$

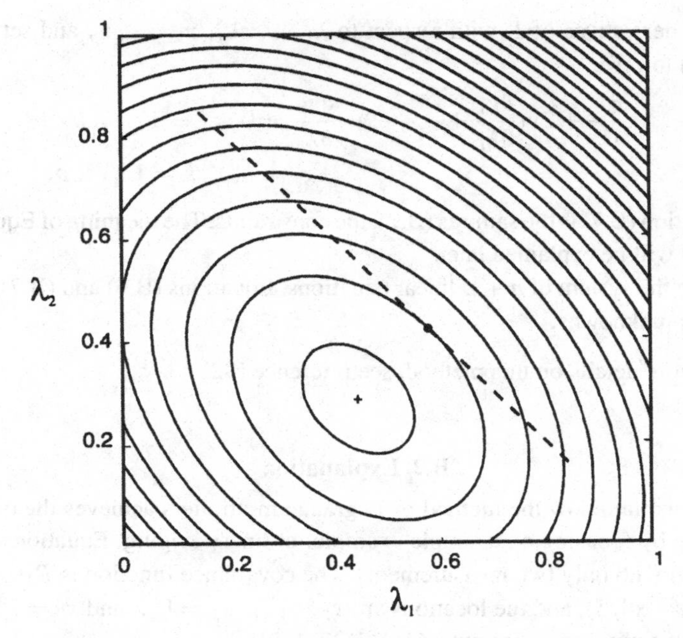

Figure B.1 Contour lines of constant MSE and the line of the constant.

computed at a point is perpendicular to the line of constant f that passes through this point. Furthermore, the vector

$$\nabla g_1 = \begin{bmatrix} \frac{\partial g_1}{\partial \lambda_1} \\ \frac{\partial g_1}{\partial \lambda_2} \end{bmatrix}$$

is perpendicular to the line representing the constraint.

The equation

$$\begin{bmatrix} \frac{\partial f}{\partial \lambda_1} \\ \frac{\partial f}{\partial \lambda_2} \end{bmatrix} + 2\nu \begin{bmatrix} \frac{\partial g_1}{\partial \lambda_1} \\ \frac{\partial g_1}{\partial \lambda_2} \end{bmatrix} = 0 \tag{B.10}$$

expresses the geometric condition that the line of constant f and the line of the constraint are parallel at the point of the optimum λs. This is the meaning of Equation (B.6).

Appendix C
Generation of realizations

In the text, we make references to the generation of realizations of random variables or functions. This appendix presents a rudimentary introduction to this subject.

C.1 Background

Generation of realizations (or sample values) of random variables and functions with given probability distributions on a digital computer is an important topic in numerical analysis [111] about which much has been written. It is useful as a numerical method for deriving probability distributions of statistics and for testing the effectiveness of estimation methodologies. In reality, the numbers generated by a digital computer (unlike the outcomes of analog devices used in lotteries) following specific algorithms only look random to someone who is not intimately familiar with the exact algorithm that is followed. The generated numbers are called pseudorandom variates and for practical purposes can be treated as random.

On a digital computer, it all starts with the generation of numbers that are uniformly distributed between 0 and 1. A simple method sets out from an integer "seed number" r_0 to generate other integers according to an equation of the type

$$r_i = (ar_{i-1} + b)\, mod(d), \quad i = 1, 2, \ldots, \qquad \text{(C.1)}$$

where a, b, and d are given integer values and $A\, mod\, B$ signifies the remainder of the division of A by B. (For example, $37\, mod 7 = 2$.) Then, the numbers $u_i = r_i/d$ are like the realizations of a random variable with uniform distribution between 0 and 1. The numbers a, b, and d vary from algorithm to algorithm and from computer to computer. For illustration, consider

$$a = 2^{24} + 9, \quad b = 1, \quad d = 2^{35}. \qquad \text{(C.2)}$$

Given a source of uniform pseudorandom variates, one can generate through transformations variates that follow the normal distribution with mean 0 and variance 1,

$$p(\eta) = \frac{1}{\sqrt{2\pi}} \exp(-\eta^2/2). \tag{C.3}$$

A common transformation is the Box-Muller. From two uniform variates (u_1, u_2) we generate two normal variates through the equations

$$\begin{aligned} \eta_1 &= \sqrt{-2\ln u_1}\cos(2\pi u_2), \\ \eta_2 &= \sqrt{-2\ln u_1}\sin(2\pi u_2). \end{aligned} \tag{C.4}$$

In this appendix, we will assume that we have a source of standard normal variates, *i.e.*, realizations of Gaussian random variables with mean 0 and variance 1, generated independently of each other. Practically all computer systems or packages of scientific software contain procedures for generation of realizations of η. These realizations are known as "normal variates." In MATLAB 4, function `rand` generates uniformly distributed variates; to generate N variates, use the following command:

```
x = rand(N,1)
```

Function `randn` has the same capacity for normal variates; to generate, for example, a matrix of $N \times M$ normal variates, just type:

```
X = randn(N,M)
```

Exercise C.1 *For practice, generate 1,000 normal variates, (a) plot their histogram, and (b) compute their arithmetic mean and variance and check whether they are close enough to the values you expected.*

C.2 Linear transformation

We will limit our focus to Gaussian random variables and processes. The most important property of interest here is that the linear transformation of normal random variables gives also Gaussian variables.

In applications, we often have to deal with the following tasks:

1. The generation of a random vector with given first and second moment. For example, this is needed in testing the significance of some statistics used in diagnostic checking. We will review a general method.
2. The generation of realizations of a stationary or stationary-increment random field with given covariance function. For example, we can have realizations of a stationary random field with mean 1 and exponential covariance with variance 1 and scale parameter 2.

C.3 Random vector

Let $\mathbf{y}$ be an $n \times 1$ Gaussian random vector with mean $\mathbf{m}$ and covariance matrix $\mathbf{Q}$. That is, y_i is a random variable, m_i is the mean of y_i and Q_{ij} is the covariance of y_i and y_j, for $i, j = 1, \ldots, n$. Then, we can obtain realizations of this vector from the following equation:

$$\mathbf{y} = \mathbf{m} + \mathbf{B}\mathbf{u}, \tag{C.5}$$

where $\mathbf{u}$ is an $n \times 1$ vector of standard normal variates and $\mathbf{B}$ is an $n \times n$ matrix such that

$$\mathbf{B}\mathbf{B}^T = \mathbf{Q}. \tag{C.6}$$

Proof $\mathbf{y}$ is Gaussian as a linear function of Gaussian $\mathbf{u}$. Its mean value can be found by taking expected values $E[\mathbf{y}] = \mathbf{m} + \mathbf{B}E[\mathbf{u}] = \mathbf{m}$. Its covariance matrix is $E[(\mathbf{y} - \mathbf{m})(\mathbf{y} - \mathbf{m})^T] = E[\mathbf{B}\,\mathbf{u}\,\mathbf{u}^T\mathbf{B}^T] = \mathbf{B}E[\mathbf{u}\,\mathbf{u}^T]\mathbf{B}^T = \mathbf{B}\mathbf{B}^T = \mathbf{Q}$.

The question now is how to find $\mathbf{B}$, known as a square root of $\mathbf{Q}$. Actually, there are multiple square roots of a matrix, in the same way that 2 and -2 are both square roots of 4. Our algorithm requires that we compute any one of them. A simple and efficient way to compute the root is through the Lower Triangular Cholesky decomposition, to be described later. We start by generating a vector with zero mean and given covariance matrix. Adding the mean is trivial. See reference [4] for additional information. For example,[1] using MATLAB:

```
Q = zeros(3,3) %initialize
for i=1:3, j=1:3 %open ''do'' loops
Q(i,j) = exp(-abs(i-j)); %covariance matrix
end, end %end do loops
B = (chol(Q))' %cholesky decomposition, then take
       %transpose
y = B*randn(3,1) %find zero-mean y
```

Let us illustrate how the Cholesky decomposition algorithm works through a simple example.

Example C.1 *Let*

$$\mathbf{Q} = \begin{bmatrix} 1 & 0.8 \\ 0.8 & 1 \end{bmatrix}.$$

Then,

$$\mathbf{B}\mathbf{B}^{\mathrm{T}} = \begin{bmatrix} B_{11}^2 & B_{11}B_{21} \\ B_{11}B_{21} & B_{21}^2 + B_{22}^2 \end{bmatrix},$$

[1] The method below applies only if $\mathbf{Q}$ is positive definite. Extensions for algorithmically singular or for conditionally positive definite matrices are beyond our scope.

where $B_{11} = 1$, $B_{12} = 0.8$, *and* $B_{22} = \sqrt{1 - 0.8^2} = 0.6$; *hence,*

$$\mathbf{B} = \begin{bmatrix} 1 & 0 \\ 0.8 & 0.6 \end{bmatrix}.$$

Exercise C.2 *Generate realizations of a Gaussian vector with mean vector*

$$\begin{bmatrix} 1 \\ 2 \\ 3 \end{bmatrix}$$

and covariance matrix

$$\begin{bmatrix} 1 & e^{-1} & e^{-2} \\ e^{-1} & 1 & e^{-1} \\ e^{-2} & e^{-1} & 1 \end{bmatrix}.$$

C.4 Stationary random functions

The method of the previous section is general but not necessarily efficient computationally. The computational cost increases with the number of elements of $\mathbf{y}$ raised to a power near 3. However, for the special case of generating realizations of a stationary (or stationary-increment) process on a regular grid, there are algorithms that are much faster; they are based on the decomposition of the power spectrum (the Fourier transform of the covariance function.)

The topic is technical and requires some understanding of power spectra and fast Fourier transforms. It is beyond the scope of this work. Useful references include [13, 48, 61, 93, 118, and 135].

References

[1] Abourifassi, M. and M. A. Marino, Cokriging of aquifer transmissivities from field measurements of transmissivity and specific capacity, *Math. Geology*, 16(1), 19–35, 1984.

[2] Abramowitz, M. and I. A. Stegun, *Handbook of Mathematical Functions*, Dover, 1965.

[3] Ahmed, S. and G. deMarsily, Comparison of geostatistical methods for estimating transmissivity using data on transmissivity and specific capacity, *Water Resources Res.*, 23(9), 1717–37, 1987.

[4] Alabert, F., The practice of fast conditional simulations through the LU decomposition of the covariance matrix, *Math. Geology*, 19(5), 369–86, 1987.

[5] Armstrong, M., Improving the Estimation and Modeling of the Variogram, in *Geostatistics for Natural Resources Characterization*, Verly et al. (eds.), Vol. 1, 1–20, Reidel, 1984.

[6] Armstrong, M., Common problems seen in variograms, *Math. Geology*, 16(3), 305–13, 1984.

[7] ASCE Task Committee on Geostatistical Techniques in Geohydrology, Review of geostatistics in geohydrology, Parts I and II, *ASCE J. Hydraulic Engineering*, 116(5), 612–58, 1990.

[8] Bakr, A., L. W. Gelhar, A. L. Gutjahr, and J. R. MacMillan, Stochastic analysis of spatial variability of subsurface flows, 1. Comparison of one- and three-dimensional flows, *Water Resources Res.*, 14(2), 263–71, 1978.

[9] Belsley, D. A., E. Kuh, and R. E. Welsch, *Regression Diagnostics*, Wiley, New York, 1980.

[10] Benjamin, J. R. and C. A. Cornell, *Probability, Statistics, and Decision for Civil Engineers*, McGraw-Hill, 1970.

[11] Borgman, L. E., New advances in methodology for statistical tests useful in geostatistical studies, *Math. Geology*, 20(4), 383–403, 1988.

[12] Box, G. E. P. and G. M. Jenkins, *Time Series Analysis*, Holden-Day, San Francisco, 1976.

[13] Bras, R. L. and I. Rodriguez-Iturbe, *Random Functions in Hydrology*, Addison-Wesley, 1985.

[14] Brooker, P. I., A parametric study of robustness of kriging variance as a function of range and relative nugget effect for a spherical semivariogram, *Math. Geology*, 18(5), 477–88, 1986.

[15] Burrough, P. A., J. Bouma, and S. R. Yates, The state of art in pedometrics, *Geoderma*, 62, 311–26, 1994.

[16] Campell, C., Bootstrapped methods for intrinsic random functions, *Math. Geology*, 20(6), 699–715, 1988.
[17] Carrera, J. and S. P. Neuman, Estimation of aquifer parameters under transient and steady state conditions, 1. Maximum likelihood method incorporating prior information, *Water Resources Res.*, 22(2), 199–210, 1986.
[18] Carrera, J. and S. P. Neuman, Estimation of aquifer parameters under transient and steady state conditions, 2. Uniqueness, stability, and solution algorithms, *Water Resources Res.*, 22(2), 211–27, 1986.
[19] Carrera, J. and L. Glorioso, On geostatistical formulations of the groundwater flow inverse problem, *Adv. Water Resources*, 14(5), 273–83, 1991.
[20] Chambers, J. M., W. S. Cleveland, B. Kleiner, and P. A. Tuckey, *Graphical Methods for Data Analysis,* Wadsworth, Belmont, CA, 395 pp., 1983.
[21] Christakos, G., On the problem of permissible covariance and variogram models, *Water Resources Res.*, 20(2), 251–65, 1984.
[22] Christakos, G., Optimal estimation of nonlinear state nonlinear observation systems, *J. Optimization Theory and Applications*, 62(1), 29–48, 1988.
[23] Christensen, R., The equivalence of predictions from universal kriging and intrinsic random-function kriging, *Math. Geology*, 22(6), 655–64, 1990.
[24] Clark, I., *Practical Geostatistics*, Applied Science Pub., London, 129 pp., 1979.
[25] Cooley, R. L., Incorporation of prior information on parameters into nonlinear regression groundwater flow models, 1. Theory, *Water Resources Res.*, 18(4), 965–7, 1982.
[26] Cooley, R. L., Incorporation of prior information on parameters into nonlinear regression groundwater flow models, 2. Applications, *Water Resources Res.*, 19(3), 662–76, 1983.
[27] Copty, N., Y. Rubin, and G. Mavko, Geophysical-hydrological identification of field permeabilities through Bayesian updating, *Water Resources Res.*, 29(8), 2813–25, 1993.
[28] Corbeil, R. R. and S. R. Searle, Restricted maximum likelihood estimation of variance components in the mixed model, *Technometrics*, 18(1), 31–38, 1976.
[29] Cressie, N., Kriging nonstationary data, *J. Amer. Statistical Assoc.*, 81, 625–34, 1986.
[30] Cressie, N. A. C., *Statistics for Spatial Data*, Wiley, New York, 900 pp., 1993.
[31] Cressie, N. and D. Hawkins, Robust estimation of the variogram, *J. Math. Geology*, 12(2), 115–26, 1980.
[32] Dagan, G., Stochastic modeling of groundwater flow by unconditional and conditional probabilities, 1. Conditional simulation and the direct problem, *Water Resources Res.*, 18(4), 813–33, 1982.
[33] Dagan, G., Stochastic modeling of groundwater flow by unconditional and conditional probabilities. The inverse problem, *Water Resources Res.*, 21(1), 65–72, 1985.
[34] Dagan, G., A note on higher-order corrections of the head covariances in steady aquifer flow, *Water Resources Res.*, 21(4), 573–8, 1985.
[35] Dagan, G., *Flow and Transport in Porous Media*, Springer-Verlag, Berlin, 465 pp., 1989.
[36] David, M., *Geostatistical Ore Reserve Estimation*, Elsevier, Amsterdam, 364 pp., 1977.
[37] David, M., *Handbook of Applied Advanced Geostatistical Ore Reserve Estimation*, Elsevier, Amsterdam, 216 pp., 1988.

[38] Delfiner, P., Linear estimation of nonstationary spatial phenomena, in *Advanced Geostatistics in the Mining Industry*, M. Guarascio, M. David, and C. Huijbregts (eds.), pp. 49–68, Reidel, Bingham, MA, 1976.
[39] Delhomme, J. P., Kriging in the hydrosciences, *Adv. Water Resources*, 1(5), 251–66, 1978.
[40] Delhomme, J. P., Spatial variability and uncertainty in groundwater flow parameters: A geostatistical approach, *Water Resources Res.*, 15(2), 269–80, 1989.
[41] deMarsily, G., *Quantitative Hydrogeology*, Academic Press, New York, 440 pp., 1986.
[42] Desbarats, A. J. and R. M. Srivastava, Geostatistical simulation of groundwater flow parameters in a simulated aquifer, *Water Resources Res.*, 27(5), 687–98, 1991.
[43] Deutsch, C. V. and A. G. Journel, *GSLIB: Geostatistical Software Library and User's Guide*, Oxford Univ. Press, New York, 1992.
[44] Dimitrakopoulos, R. (ed.), *Geostatistics for the Next Century: An International Forum in Honour of Michel David's Contribution to Geostatistics*, Kluwer, Dordrecht, Boston, 497 pp., 1994.
[45] Drake, A. W., *Fundamentals of Applied Probability Theory*, McGraw-Hill, 1967.
[46] Dubrule, O., Cross validation of kriging in a unique neighborhood, *Math. Geology*, 15(6), 687–99, 1983.
[47] Dubrule, O., Two methods with different objectives: Splines and kriging, *Math. Geology*, 15(2), 245–57, 1983.
[48] Dykaar, B. B. and P. K. Kitanidis, Determination of the effective hydraulic conductivity for heterogeneous porous media using a numerical spectral approach, 1. Method, *Water Resources Res.*, 28(4), 1155–66, 1992.
[49] Emsellem, Y. and G. deMarsily, An automatic solution for the inverse problem, *Water Resources Res.*, 7(5), 1264–83, 1971.
[50] Englund, E. and A. Sparks, GEO-EAS (Geostatistical Environmental Assessment Software) User's Guide, *U.S. EPA* 600/4-88/033a, 1988.
[51] Feinerman, E., G. Dagan, and E. Bresler, Statistical inference of spatial random functions, *Water Resources Res.*, 22(6), 935–42, 1986.
[52] Filliben, J. J., The probability plot correlation test for normality, *Technometrics*, 17(1), 111–17, 1975.
[53] Freeze, R. A., A stochastic-conceptual analysis of one-dimensional groundwater flow in nonuniform homogeneous media, *Water Resources Res.*, 11(5), 725–41, 1975.
[54] Gambolati, G. and G. Volpi, Groundwater contour mapping in Venice by stochastic interpolators, 1—Theory, *Water Resources Res.*, 15(2), 281–90, 1979.
[55] Gandin, L. S., *Objective Analysis of Meteological Fields*, Israel Program for Scientific Translations (available from the U.S. Dept. of Commerce, Clearinghouse for Federal Scientific and Technical Information, Springfield, VA), Jerusalem, 242 pp., 1965.
[56] Gavalas, G. R., P. C. Shah, and J. H. Seinfeld, Reservoir history matching by Bayesian estimation, *Soc. Petrol. Eng. J.*, 337–50, 1976.
[57] Gelhar, L. W., *Stochastic Subsurface Hydrology*, Prentice Hall, Englewood Cliffs, NJ, 1993.
[58] Gelhar, L. W. and C. L. Axness, Three-dimensional stochastic analysis of macrodispersion, *Water Resources Res.*, 19(1), 161–80, 1983.

[59] Graham, W. and D. McLaughlin, Stochastic analysis of nonstationary subsurface solute transport, 2. Conditional moments, *Water Resources Res.*, 25(11), 2331–55, 1989.
[60] Gutjahr, A. L. and L. W. Gelhar, Stochastic models of subsurface flows: Infinite versus finite domains and stationarity, *Water Resources Res.*, 17(2), 337–51, 1981.
[61] Gutjahr, A., B. Bullard, S. Hatch, and L. Hughson, Joint conditional simulations and the spectral approach for flow modeling, *Stoch. Hydrol. Hydraul.*, 8(1), 79–108, 1994.
[62] Hald, A., *Statistical Tables and Formulas*, Wiley, New York, 1952.
[63] Hardy, R. L., Theory and applications of the multiquadric-biharmonic method: 20 years of discovery, *Computers Math. Applic.*, 19(8/9), 163–208, 1990.
[64] Hoeksema, R. J. and P. K. Kitanidis, An application of the geostatistical approach to the inverse problem in two-dimensional groundwater modeling, *Water Resources Res.*, 20(7), 1003–20, 1984.
[65] Hoeksema, R. J. and P. K. Kitanidis, Comparison of Gaussian conditional mean and kriging estimation in the geostatistical solution of the inverse problem, *Water Resources Res.*, 21(6), 825–36, 1985.
[66] Hoeksema, R. J. and P. K. Kitanidis, Analysis of the spatial structure of properties of selected aquifers, *Water Resources Res.*, 21(4), 563–72, 1985.
[67] Hoeksema, R. J. and P. K. Kitanidis, Prediction of transmissivities, heads, and seepage velocities using mathematical models and geostatistics, *Adv. Water Resources*, 12(2), 90–102, 1989.
[68] Hoeksema, R. J., R. B. Clapp, A. L. Thomas, A. E. Hunley, N. D. Farrow, and K. C. Dearstone, Cokriging model for estimation of water table elevation, *Water Resources Res.*, 25(3), 429–38, 1989.
[69] Hoeksema, R. J. and R. B. Clapp, Calibration of groundwater flow models using Monte Carlo simulations and geostatistics, in *Model CARE 90: Calibration and Reliability in Groundwater Modelling* (pp. 33–42). IAHS Publ. No. 195, 1990.
[70] Hohn, M. E., *Geostatistics and Petroleum Geology*, Van Nostrand Reinhold, New York, 264 pp., 1988.
[71] Hughes, J. P., Data requirements for kriging: Estimation and network design, *Water Resources Res.*, 17(6), 1641–50, 1981.
[72] Huyakorn, P. S. and G. F. Pinder, *Computational Methods in Subsurface Flow*, Academic Press, New York, 1983.
[73] Isaaks, E. H. and R. M. Srivastava, *Applied Geostatistics*, Oxford Univ. Press, New York, 561 pp., 1989.
[74] Jahns, H. O., A rapid method for obtaining a two-dimensional reservoir description from well pressure response data, *Soc. Petrol. Eng. J.*, 315–27, 1966.
[75] Journel, A., The place of non-parametric geostatistics, in *Geostatistics for Natural Resources Characterization*, Verly et al. (eds), Reidel, Dordrecht, The Netherlands, 1984.
[76] Journel, A. G. and Ch. J. Huigbregts, *Mining Geostatistics*, Academic Press, 600 pp., 1978.
[77] Kafritsas, J. and R. L. Bras, *The Practice of Kriging*, R. L. Parsons Lab., M.I.T, 1981.
[78] Kimeldorf, G. and G. Wahba, A correspondence between Bayesian estimation of stochastic processes and smoothing by splines, *Ann. Math. Statistics*, 41, 495–502, 1970.

[79] Kitanidis, P. K., Statistical estimation of polynomial generalized covariance functions and hydrologic applications, *Water Resources Res.*, 19(4), 909–21, 1983.

[80] Kitanidis, P. K., Parameter uncertainty in estimation of spatial functions: Bayesian analysis, *Water Resources Res.*, 22(4), 499–507, 1986.

[81] Kitanidis, P. K., Parameter estimation of covariances of regionalized variables, *Water Resources Bulletin*, 23(4), 557–67, 1987.

[82] Kitanidis, P. K., Orthonormal residuals in geostatistics: Model criticism and parameter estimation, *Math. Geology*, 23(5), 741–58, 1991.

[83] Kitanidis, P. K., generalized covariance functions in estimation, *Math. Geology*, 25(5), 525–40, 1993.

[84] Kitanidis, P. K., Geostatistics, in *Handbook of Hydrology*, D. R. Maidment (ed.), MacGraw-Hill, New York, 20.1–20.39, 1993.

[85] Kitanidis, P. K., Quasilinear geostatistical theory for inversing, *Water Resources Res.*, 31(10), 2411–19, 1995.

[86] Kitanidis, P. K., Analytical expressions of conditional mean, covariance, and sample functions in geostatistics, *Stoch. Hydrol. Hydraulics*, 10(4), 279–294, 1996.

[87] Kitanidis, P. K. and E. G. Vomvoris, A geostatistical approach to the inverse problem in groundwater modeling (steady state) and one-dimensional simulations, *Water Resources Res.*, 19(3), 677–90, 1983.

[88] Kitanidis, P. K. and R. W. Lane, Maximum likelihood parameter estimation of hydrologic spatial processes by the Gauss-Newton method, *J. Hydrology*, 79, 53–79, 1985.

[89] Krajewski, W. F., Co-kriging of radar-rainfall and rain gage data, *J. Geophys. Res.*, 92(D8), 9571–80, 1987.

[90] Linsley, R. K., M. A. Kohler, and J. L. H. Paulhus, *Hydrology for Engineers*, McGraw-Hill, New York, 1982.

[91] Loaiciga, H. A. and M. A. Marino, On solution of the inverse problem for confined aquifer flow via maximum likelihood, *Math. Geology*, 18(7), 677–92, 1986.

[92] Luenberger, D. G., *Introduction to Linear and Nonlinear Programming*, Addison-Wesley, Reading, MA, 1973.

[93] Mantoglou, A. and J. L. Wilson, The turning bands method for simulation of random fields using line generation by a spectral method, *Water Resources Res.*, 18(5), 1379–94, 1982.

[94] Matheron, G., Principles of geostatistics, *Economic Geology*, 58, 1246–66, 1963.

[95] Matheron, G., *The Theory of Regionalized Variables and Its Applications*, Ecole de Mines, Fontainbleau, France, 212 pp., 1971.

[96] Matheron, G., The intrinsic random functions and their applications, *Adv. Appl. Prob.*, 5, 439–68, 1973.

[97] Matheron, G., Splines and kriging: Their formal equivalence, in *Down-to-Earth Statistics: Solutions Looking for Geological Problems*, D. F. Merriam (ed.), pp. 77–95, Syracuse Univ. Geological Contributions, Syracuse, NY, 1981.

[98] Matheron, G. and M. Armstrong, Disjunctive kriging revisited, parts I and II, *Math. Geology*, 18(8), 711–42, 1986.

[99] Micchelli, C. A., Interpolation of scattered data: Distance matrices and conditionally positive definite functions, *Constructive Approximation*, 2, 11–22, 1986.

[100] Myers, D. E., Matrix formulation of cokriging, *J. Math. Geology*, 14(3), 249–57, 1982.

[101] Neuman, S. P., A statistical approach to the inverse problem of aquifer hydrology, 3. Improved method and added perspectives, *Water Resources Res.*, 16(2), 331–46, 1980.

[102] Neuman, S. P., Role of geostatistics in subsurface hydrology, in *Geostatistics for Natural Resources Characterization*, Verly et al. (eds.), Part I, pp. 787–816, Reidel, Dordrecht, The Netherlands, 1984.

[103] Neuman, S. P., G. E. Fogg, and E. A. Jacobson, A statistical approach to the inverse problem of aquifer hydrology, II. Case study, *Water Resources Res.*, 16(1), 33–58, 1980.

[104] Neuman, S. P. and E. A. Jacobson, Analysis of nonintrinsic spatial variability by residual kriging with applications to regional groundwater levels, *Math. Geology*, 16(5), 499–521, 1984.

[105] Olea, R. A., Measuring spatial dependence with semivariographs, Report, Kansas Geol. Survey, 1975.

[106] Olea, R. A., Optimum mapping techniques using regionalized variable theory, Report, Kansas Geol. Survey, 1975.

[107] Omre, E., The variogram and its estimation, in *Geostatistics for Natural Resources Characterization*, Verly et al. (eds.), Vol. 1, pp. 107–25, D. Reidel, Dordrecht, The Netherlands, 1984.

[108] Omre, H., Bayesian kriging—Merging observations and qualified guesses in kriging, *Math. Geology*, 19(1), 25–39, 1987.

[109] Pearson, E. S. and H. O. Hartley, *Biometrika Tables for Statisticians*, Cambridge Univ. Press, New York, 1954.

[110] Philip, R. D. and P. K. Kitanidis, Geostatistical estimation of hydraulic head gradients, *Ground Water*, 27(6), 855–65, 1989.

[111] Press, W. H., B. P. Flannery, S. A. Teukolsky, and W. T. Vetterling, *Numerical Recipes in C*, Cambridge Univ. Press, Cambridge, England, 735 pp., 1988.

[112] Puente, C. E. and R. L. Bras, Disjunctive kriging, universal kringing, or no kriging: Small sample results with simulated fields, *Math. Geology*, 18(3), 287–305, 1986.

[113] Rao, C. R., *Linear Statistical Inference and Its Applications*, Wiley, New York, 625 pp., 1973.

[114] Rendu, J. M., Disjunctive kriging: A simplified theory, *Math. Geology*, 12, 306–21, 1980.

[115] Rendu, J., Disjunctive kriging: Comparison of theory with actual results, *Math. Geology*, 12(4), 305–20, 1980.

[116] Ritzi, R. W., D. F. Jayne, A. J. Zahradnik, A. A. Field, and G. E. Fogg, Geostatistical modeling of heterogeneity in glaciofluvial, buried-valley aquifers, *Ground Water*, 32(4), 666–74, 1994.

[117] Rivoirard, J., *Introduction to Disjunctive Kriging and Non-Linear Geostatistics*, Oxford Univ. Press, New York, 1994.

[118] Robin, M., A. Gutjahr, E. Sudicky, and J. L. Wilson, Cross-correlated random field generation with the direct Fourier transform method, *Water Resources Res.*, 29(7), 2385–97, 1993.

[119] Rouhani, S., Comparative study of groundwater mapping techniques, *J. Ground Water*, 24(2), 207–16, 1986.

[120] Rouhani, S. and T. J. Hall, Geostatistical schemes for groundwater sampling, *J. Hydrology*, 103, 85–102, 1988.

[121] Royle, A. G. et al., *Geostatistics*, McGraw-Hill, New York, 168 pp., 1980.

[122] Rubin, Y. and G. Dagan, Stochastic identification of transmissivity and effective recharge in steady groundwater flow, 1. Theory, *Water Resources Res.*, 23(7), 1185–92, 1987.

[123] Rubin, Y. and G. Dagan, Stochastic identification of transmissivity and effective recharge in steady groundwater flow, 2. Case study, *Water Resources Res.*, 23(7), 1193–200, 1987.

[124] Schweppe, F. C., *Uncertain Dynamic Systems*, Prentice-Hall, Englewood-Cliffs, NJ, 1973.

[125] Shah, P. C., G. R. Gavalas, and J. H. Seinfeld, Error analysis in history matching: The optimum level of parametrization, *Soc. Petrol. Eng. J.*, 219–28, 1978.

[126] Shapiro, S. S. and M. B. Wilk, An analysis of variance test for normality, (complete samples), *Biometrika*, 52, 691–710, 1965.

[127] Smith, L. and R. A. Freeze, Stochastic analysis of steady state groundwater flow in a bounded domain, 1. One-dimensional simulations, *Water Resources Res.*, 15(3), 521–28, 1979.

[128] Smith, L. and R. A. Freeze, Stochastic analysis of steady state groundwater flow in a bounded domain, 2. Two-dimensional simulations, *Water Resources Res.*, 15(6), 1543–59, 1979.

[129] Starks, T. H. and J. H. Fang, The effect of drift on the experimental semivariogram, *Math. Geology*, 14(4), 309–19, 1982.

[130] Stein, M. L., Minimum norm quadratic estimation of spatial variograms, *J. Amer. Statistical Assoc.*, 82(399), 765–72, 1987.

[131] Strang, G., *Linear Algebra and Its Applications*, Harcourt Brace Jovanovich, San Diego, 505 pp., 1988.

[132] Sullivan, J., Conditional recovery estimation through probability kriging—Theory and practice, in *Geostatistics for Natural Resources Characterization*, Verly et al. (eds.), Reidel, Dordrecht, The Netherlands, 1984.

[133] Sun, N.-Z., *Inverse Problems in Groundwater Modeling*, Kluwer, Norwell, MA, 352 pp., 1994.

[134] Switzer, P. and H. M. Parker, The problem of ore versus waste discrimination for individual blocks: The lognormal model, in *Advanced Geostatistics in the Mining Industry*, M. Guarascio et al. (eds.), pp. 203–18, Reidel, Dordrecht, The Netherlands, 1976.

[135] Tompson, A. F. B., R. Ababou, and L. W. Gelhar, Implementation of the three-dimensional turning bands random field generator, *Water Resources Res.*, 25(10), 2227–43, 1989.

[136] Vauclin, M., S. R. Vieira, G. Vachaud, and D. R. Nielsen, The use of cokriging with limited field soil information, *Soil Sci. Proc. Amer. J.*, 47(2), 175–84, 1983.

[137] Volpi, G. and G. Gambolati, On the use of a main trend for the kriging technique in hydrology, *Adv. Water Resources*, 1(6), 345–9, 1978.

[138] Wagner, B. J. and S. M. Gorelick, Reliable aquifer remediation in the presence of spatially variable hydraulic conductivity: From data to design, *Water Resources Res.*, 25(10), 2211–25, 1989.

[139] Wang, H. and M. P. Anderson, *Introduction to Groundwater Modeling: Finite Differences and Finite Element Methods*, Freeman, San Francisco, 1982.

[140] Warnes, J. J., A sensitivity analysis for universal kriging, *Math. Geology*, 18(7), 653–76, 1986.

[141] Warrick, A. W., D. E. Myers, and D. R. Nielsen, Geostatistical methods applied to soil sciences, in *Methods of Soil Analysis, Part 1, Agronomy Monogram No. 9*, Amer. Soc. Agronomy, Madison, Wis., 1986.

[142] Watson, G. S., Smoothing and interpolation by kriging and with splines, *Math. Geology*, 16(6), 601–15, 1984.

[143] Yates, S. R., A. W. Warrick, and D. E. Myers, Disjunctive kriging Parts I and II. Overview of estimation and conditional probability, *Water Resources Res.*, 22, 615–30, 1986.
[144] Yates, S. R. and A. W. Warrick, Estimating soil water content using cokriging, *Soil Sci. Soc. Amer.*, 51, 23–30, 1987.
[145] Yates, S. R. and M. V. Yates, *Geostatistics for Waste Management: User's Manual for GEOPACK Version 1.0*, Kerr Environmental Research, 1989.
[146] Yeh, W. W.-G. and Y. S. Yoon, Aquifer parameter identifiability with optimum dimension in parameterization, *Water Resources Res.*, 17(3), 664–72, 1981.
[147] Yeh, W. W.-G., Review of parameter identification procedures in groundwater hydrology: The inverse problem, *Water Resources Res.*, 22(1), 95–108, 1986.
[148] Yeh, J. T.-C., M. Jin, and S. Hanna, An iterative stochastic inverse method: Conditional effective transmissivity and hydraulic head fields, *Water Resources Res.*, 32(1), 85–92, 1996.
[149] Zimmerman, D. L. and M. B. Zimmerman, A comparison of spatial semivariogram estimators and corresponding ordinary kriging predictors, *Technometrics*, 33(1), 77–91, 1991.

Index

Printed in the United States
By Bookmasters